Rhondda-Cynon-Taf County Borough Council
Llyfrgelloedd Bwrdeistref Sirol Rhondda-Cynon Taf

This book is available for consultation on the premises only
Mae'r llyfr hwn ar gael i'w ymgynghori ofewn yr adeilad hwn yn unig

HQ
RGF **B**

Contents

PAGE

USE OF GUIDANCE   5

General introduction: Fire safety   8

**MEANS OF WARNING AND ESCAPE
THE REQUIREMENT B1**   11

GUIDANCE   12
Performance   12
Introduction   12
Interaction with other legislation   12
Management of premises   13
Analysis of the problem   13
Means of escape for disabled people   14
Security   14
Alternative approaches   14
Use of the document   15
Methods of measurement   15

**Section 1: Fire alarm and fire
detection systems**   17
Introduction   17
Dwellings   17
Buildings other than dwellings   18

Section 2: Dwellinghouses   21
Introduction   21
General provisions   21
Provisions for escape from floors not
more than 4.5m above ground level   21
Additional provisions for houses with
a floor more than 4.5m above ground level   22
Loft conversions   24

Section 3: Flats and maisonettes   27
Introduction   27
General provisions   27
Provisions for escape from flats and
maisonettes where the floor is not more
than 4.5m above ground level   28
Additional provisions for flats and
maisonettes with a floor more than
4.5m above ground level   28
Means of escape in the common parts
of flats and maisonettes   31
Number of escape routes   31
Common escape routes   33
Common stairs   34
Protection of common stairs   34
Basement stairs   35
Stairs serving accommodation ancillary
to flats and maisonettes   35
External escape stairs   35
Dwellings in mixed use buildings   35

PAGE

**Section 4: Design for horizontal escape
– buildings other than dwellings**   36
Introduction   36
Escape route design   36
Hospitals and other residential care
premises of Purpose Group 2a   42

**Section 5: Design for vertical escape
– buildings other than dwellings**   43
Introduction   43
Number of escape stairs   43
Width of escape stairs   43
Calculation of minimum stair width   44
Protection of escape stairs   47
Basement stairs   48
External escape stairs   48

**Section 6: General provisions common
to buildings other than dwellinghouses**   49
Introduction   49
Protection of escape routes   49
Doors on escape routes   49
Stairs   50
General   53
Lifts   54
Mechanical ventilation and
air conditioning systems   55
Refuse chutes and storage   55
Shop store rooms   55

**INTERNAL FIRE SPREAD (LININGS)
THE REQUIREMENT B2**   56
GUIDANCE   57
Performance   57
Introduction   57

Section 7: Wall and ceiling linings   58
Classification of linings   58
Variations and special provisions   58
Thermoplastic materials   59

**INTERNAL FIRE SPREAD (STRUCTURE)
THE REQUIREMENT B3**   61
GUIDANCE   62
Performance   62
Introduction   62

**Section 8: Loadbearing elements
of structure**   63
Introduction   63
Fire resistance standard   63
Floors in domestic loft conversions   63
Raised storage areas   64
Conversion to flats   64

B

	PAGE
Section 9: Compartmentation	65
Introduction	65
Provision of compartmentation	65
Construction of compartment walls and compartment floors	68
Openings in compartmentation	70
Protected shafts	70
Section 10: Concealed spaces (cavities)	74
Introduction	74
Provision of cavity barriers	74
Construction and fixings for cavity barriers	77
Maximum dimensions of concealed spaces	78
Openings in cavity barriers	78
Section 11: Protection of openings and fire stopping	80
Introduction	80
Openings for pipes	80
Ventilating ducts	81
Flues, etc.	81
Fire-stopping	81
Section 12: Special provisions for car parks and shopping complexes	83
Introduction	83
Car Parks	83
Shopping complexes	84
EXTERNAL FIRE SPREAD THE REQUIREMENT B4	85
GUIDANCE	86
Performance	86
Introduction	86
Section 13: Construction of external walls	87
Introduction	87
Fire resistance standard	87
Portal frames	87
External surfaces	88
External wall construction	88
Section 14: Space separation	90
Introduction	90
Boundaries	91
Unprotected areas	92
Methods for calculating acceptable unprotected area	94
Section 15: Roof coverings	96
Introduction	96
Classification of performance	96
Separation distances	96

	PAGE
ACCESS AND FACILITIES FOR THE FIRE SERVICE THE REQUIREMENT B5	99
GUIDANCE	100
Performance	100
Introduction	100
Section 16: Fire Mains	101
Introduction	101
Provision of fire mains	101
Number and location of fire mains	101
Design and construction of fire mains	101
Section 17: Vehicle access	102
Introduction	102
Buildings not fitted with fire mains	102
Buildings fitted with fire mains	105
Design of access routes and hard standings	105
Section 18: Access to buildings for firefighting personnel	106
Introduction	106
Provision of firefighting shafts	106
Number and location of firefighting shafts	107
Design and construction of firefighting shafts	107
Rolling shutters in compartment walls	107
Section 19: Venting of heat and smoke from basements	108
Introduction	108
Provision of smoke outlets	108
Construction of outlet ducts or shafts	109
Basement car parks	109
APPENDICES	
Appendix A: Performance of materials and structures	110
Introduction	110
Fire resistance	110
Roofs	111
Reaction to fire	111
Non-combustible materials	111
Materials of limited combustibility	111
Internal linings	111
Thermoplastic materials	112
Fire test methods	112
Appendix B: Fire doors	124
Appendix C: Methods of measurement	128
Appendix D: Purpose groups	130
Appendix E: Definitions	132
Appendix F: Fire behaviour of insulating core panels used for internal structures	136
Appendix G: Standards referred to	138
Other publications referred to	142
Index	144

B

D1806367

PAGE

B1

1. Measurement of door width — 16
2. Ground or basement storey exit into an enclosed space — 22
3. Alternative arrangements for final exits — 23
4. Fire separation in houses with more than one floor over 4.5m above ground level — 23
5. Alternatives for the fire separation of the stair and new storey in house conversion — 25
6. Position of dormer window or rooflight suitable for emergency egress purposes from a loft conversion of a 2-storey dwellinghouse — 26
7. Flat where all habitable rooms have direct access to an entrance hall — 28
8. Flat with restricted travel distance from furthest point to entrance — 28
9. Flat with an alternative exit, but where all habitable rooms have no direct access to an entrance hall — 29
10. Maisonette with alternative exits from each habitable room, except at entrance level — 29
11. Maisonette with protected entrance hall and landing — 30
12. Flats or maisonettes served by one common stair — 31
13. Flats or maisonettes served by more than one common stair — 32
14. Common escape route in small single stair building — 32
15. Travel distance in dead-end condition — 38
16. Alternative escape routes — 38
17. Inner room and access room — 39
18. Exits in a central core — 39
19. Dead-end corridors — 41
20. Progressive horizontal evacuation — 42
21. External protection to protected stairways — 51
22. Fire resistance of areas adjacent to external stairs — 52

B2

23. Lighting diffuser in relation to ceiling — 59
24. Layout restrictions on Class 3 plastics rooflights, TP(b) rooflights and TP(b) lighting diffusers — 60

B3

25. Separation between garage and dwelling house — 66
26. Compartment floors: illustration of guidance in paragraph 9.20 — 67
27. Compartment walls and compartment floors with reference to relevant paragraphs in Section 9 — 69
28. Junction of compartment wall with roof — 71

PAGE

29. Protected shafts — 72
30. Uninsulated glazed screen separating protected shaft from lobby or corridor — 73
31. Interrupting concealed spaces (cavities) — 74
32. Cavity walls excluded from provisions for cavity barriers — 76
33. Alternative arrangements in roof space over protected stairway in a house with a floor more than 4.5m above ground level — 76
34. Corridor enclosure alternatives — 77
35. Fire-resisting ceiling below concealed space — 79
36. Provisions for cavity barriers in double-skinned insulated roof sheeting — 79
37. Pipes penetrating structure — 81
38. Enclosure for drainage or water supply pipes — 82
39. Flues penetrating compartment walls or floors — 82

B4

40. Provisions for external surfaces of walls — 89
41. Relevant boundary — 90
42. Notional boundary — 91
43. Status of combustible surface material as unprotected area — 92
44. Unprotected areas which may be disregarded in assessing the separation distance from the boundary — 93
45. The effect of a canopy on separation distance — 93
46. Permitted unprotected areas in small residential buildings — 95
47. Limitations on spacing and size of plastics rooflights having a Class 3 or TP(b) lower surface — 96

B5

48. Example of building footprint and perimeter — 103
49. Relationship between building and hard standing/access roads for high reach fire appliances — 104
50. Turning facilities — 105
51. Provision of firefighting shafts — 106
52. Components of a firefighting shaft — 107
53. Fire-resisting construction for smoke outlet shafts — 108

APPENDIX C

C1. Cubic Capacity — 128
C2. Area — 128
C3. Height of building — 129
C4. Number of storeys — 129
C5. Height of top storey in building — 129

APPENDIX E

E1. Recessed car parking areas — 135

PAGE

TABLES

B1

1. Floor space factors — 16
2. Limitations on distance of travel in common areas of flat and maisonette buildings — 33
3. Limitations on travel distance — 37
4. Minimum number of escape routes and exits from a room, tier or storey — 38
5. Widths of escape routes and exits — 40
6. Minimum width of escape stairs — 44
7. Capacity of a stair for basements and for simultaneous evacuation of the building — 45
8. Minimum width of stairs designed for phased evacuation — 46
9. Provisions for escape lighting — 54

B2

10. Classification of linings — 58
11. Limitations applied to thermoplastic rooflights and lighting diffusers in suspended ceilings and Class 3 plastics rooflights. — 60

B3

12. Maximum dimensions of building or compartment (non-residential buildings) — 68
13. Provision of cavity barriers — 75
14. Maximum dimensions of cavities in non-domestic buildings (purpose groups 2-7) — 76
15. Maximum nominal internal diameter of pipes passing through a compartment wall/floor — 80

B4

16. Permitted unprotected areas in small buildings or compartments — 95
17. Limitations on roof coverings — 97
18. Class 3 plastics rooflights: limitations on use and boundary distance — 98
19. TP(a) and TP(b) plastics rooflights: limitations on use and boundary distance — 98

B5

20. Fire service vehicle access to buildings (excluding blocks of flats) not fitted with fire mains — 102
21. Typical fire service vehicle access route specification — 105
22. Minimum number of firefighting shafts in building fitted with sprinklers — 107

PAGE

APPENDICES

A1. Specific provisions of test for fire resistance of elements of structure etc. — 114-116
A2. Minimum periods of fire resistance — 117
A3. Limitations on fire-protecting suspended ceilings — 118
A4. Limitations on the use of uninsulated glazed elements on escape routes — 119
A5. Notional designations of roof coverings — 120
A6. Use and definitions of non-combustible materials — 121
A7. Use and definitions of materials of limited combustibility — 122
A8. Typical performance ratings of some generic materials and products — 123
B1. Provisions for fire doors — 126-127

D1. Classification of purpose groups — 131

Use of Guidance

THE APPROVED DOCUMENTS

The Building Regulations 2000 (S.I. 2000/2531), which come into operation on 1st January 2001, replace the Building Regulations 1991 (S.I. 1991/2768) and consolidate all subsequent revisions to those regulations. This document is one of a series that has been approved and issued by the Secretary of State for the purpose of providing practical guidance with respect to the requirements of Schedule 1 to and regulation 7 of the Building Regulations 2000 for England and Wales.

This document provides a consolidation of the guidance previously issued in the original 2000 edition of Approved Document B and the subsequent amendments of it which were issued in 2000 and 2002.

At the back of this document is a list of all the documents that have been approved and issued by the Secretary of State for this purpose.

The Approved Documents are intended to provide guidance for some of the more common building situations. However, there may well be alternative ways of achieving compliance with the requirements. **Thus there is no obligation to adopt any particular solution contained in an Approved Document if you prefer to meet the relevant requirement in some other way.**

Other requirements

The guidance contained in an Approved Document relates only to the particular requirements of the Regulations which that document addresses. The building work will also have to comply with the Requirements of any other relevant paragraphs in Schedule 1 to the Regulations.

There are Approved Documents which give guidance on each of the other requirements in Schedule 1 and on regulation 7.

LIMITATION ON REQUIREMENTS

In accordance with regulation 8, the requirements in Parts A to K and N of Schedule 1 to the Building Regulations do not require anything to be done except for the purpose of securing reasonable standards of health and safety for persons in or about the building.

MATERIALS AND WORKMANSHIP

Any building work which is subject to the requirements imposed by Schedule 1 of the Building Regulations should, in accordance with regulation 7, be carried out with proper materials and in a workmanlike manner.

You may show that you have complied with regulation 7 in a number of ways. These include the appropriate use of a product bearing CE marking in accordance with the Construction Products Directive (89/106/EEC)[1] as amended by the CE Marking Directive (93/68/EEC)[2] or a product complying with an appropriate technical specification (as defined in those Directives), a British Standard, or an alternative national technical specification of any state which is a contracting party to the European Economic Area which, in use, is equivalent, or a product covered by a national or European certificate issued by a European Technical Approval Issuing body, and the conditions of use are in accordance with the terms of the certificate. You will find further guidance in the Approved Document supporting regulation 7 on materials and workmanship.

Independent certification schemes

There are many UK product certification schemes. Such schemes certify compliance with the requirements of a recognised document which is appropriate to the purpose for which the material is to be used. Materials which are not so certified may still conform to a relevant standard.

Many certification bodies which approve such schemes are accredited by UKAS.

Since the fire performance of a product, component or structure is dependent upon satisfactory site installation and maintenance, independent schemes of certification and registration of installers and maintenance firms of such will provide confidence in the appropriate standard of workmanship being provided.

Technical specifications

Building Regulations are made for specific purposes: health and safety, energy conservation and the welfare and convenience of disabled people. Standards and technical approvals are relevant guidance to the extent that they relate to these considerations. However, they may also address other aspects of performance such as serviceability, or aspects which although they relate to health and safety are not covered by the Regulations.

When an Approved Document makes reference to a named standard, the relevant version of the standard is the one listed at the end of the publication. However, if this version of the standard has been revised or updated by the issuing standards body, the new version may be used as a source of guidance provided it continues to address the relevant requirements of the Regulations.

---

[1] As implemented by the Construction Products Regulations 1991 (SI 1991 No 1620)

[2] As implemented by the Construction Products (Amendment) Regulations 1994 (SI 1994 No 3051)

The appropriate use of a product which complies with a European Technical Approval as defined in the Construction Products Directive will meet the relevant requirements.

The Department intends to issue periodic amendments to its Approved Documents to reflect emerging harmonised European Standards. Where a national standard is to be replaced by a European harmonised standard, there will be a co-existence period during which either standard may be referred to. At the end of the co-existence period the national standard will be withdrawn.

The Workplace (Health, Safety and Welfare) Regulations 1992

The Workplace (Health, Safety and Welfare) Regulations 1992 contain some requirements which affect building design. The main requirements are now covered by the Building Regulations, but for further information see: *Workplace health, safety and welfare, The Workplace (Health, Safety and Welfare) Regulations 1992, Approved Code of Practice and Guidance*; The Health and Safety Commission, L24; Published by HMSO 1992; ISBN 0-11-886333-9.

The Workplace (Health, Safety and Welfare) Regulations 1992 apply to the common parts of flats and similar buildings if people such as cleaners, wardens and caretakers are employed to work in these common parts. Where the requirements of the Building Regulations that are covered by this Part do not apply to dwellings, the provisions may still be required in the situations described above in order to satisfy the Workplace Regulations.

The Construction (Health, Safety and Welfare) Regulations 1996

The purpose of this Approved Document is to provide guidance on the fire safety requirements for the completed building. It does not address the risk of fire during the construction work which is covered by the Construction (Health, Safety and Welfare) Regulations 1996. HSE have issued the following guidance on these Regulations: Construction Information Sheet No 51 *Construction fire safety*; and HSG 168 *Fire safety in construction work* (ISBN 0-7176-1332-1).

When the construction work is being carried out on a completed building which, apart from the construction site part of the building, remains occupied, the fire authority are responsible for the enforcement of the 1996 Regulations in respect of fire. Where the building is unoccupied, the Health and Safety Executive are responsible for enforcement.

The Construction Products Directive

The Construction Products Directive (CPD) is one of the "New Approach" Directives, which seek to remove technical barriers to trade within the European Economic Area (EEA) as part of the move to complete the Single Market. The EEA comprises the European Community and those states in the European Free Trade Association (other than Switzerland). The intention of the CPD is to replace existing national standards and technical approvals with a single set of European-wide technical specifications for construction products (i.e. harmonised European standards or European Technical Approvals). Any manufacturer whose products have CE marking showing that they are specified according to European technical specifications cannot have his products refused entry to EEA markets on technical grounds. In the UK, the CPD was implemented by the Construction Products Regulations, which came into force on 27 December 1991 and were amended on 1 January 1995 by the Construction Products (Amendment) Regulations 1994.

This document refers to, and utilises within its guidance, a large number of British Standards, in relation to Codes of Practice and fire test methods (typically the BS 476 series of documents). In order to facilitate harmonisation and the use of the new technical specifications and their supporting European test standards, guidance is also given on the classification of products in accordance with those standards.

Guidance is given for the appropriate use and/or specification of a product to which one or more of the following apply:

1. a product bearing CE marking in accordance with the Construction Products Directive (89/106/EEC) as amended by the CE marking Directive (93/68/EEC);

2. a product tested and classified in accordance with the European Standards (BS EN) referred to in the Commission Decision 2000/147/EC[1] and/or Commission Decision 2000/367/EC[2];

3. a product complying with an appropriate technical specification (as defined in the Directives 89/106/EC as amended by 93/68/EEC).

The implementation of the Construction Products Directive (CPD) will necessitate a time period during which national (British) Standards and European technical specifications will co-exist. This is the so-called period of co-existence. The objective of this period of co-existence is to provide for a gradual adaptation to the requirements of the CPD. It will enable producers, importers and distributors of construction products to sell stocks of products manufactured in line with the national

---

[1] Implementing Council Directive 89/106/EEC as regards the classification of the reaction to fire (2000/147/EC) performance of construction products.

[2] Implementing Council Directive 89/106/EEC as regards the classification of the resistance to fire (2000/367/EC) performance of construction products, construction works and parts thereof.

rules previously in force and have new tests carried out. The duration of the period of co-existence in relation to the European fire tests has not yet been clearly defined.

As new information becomes available and further harmonised European standards relevant to this document are published, further guidance will be made available. For example, further guidance will be necessary in the areas of roof coverings and thermoplastics.

Designation of standards

The designation of "xxxx" is used for the year referred to for standards that are not yet published. The latest version of any standard may be used provided that it continues to address the relevant requirements of the Regulations.

Commission guidance papers and decisions

The following guidance papers and Commission decisions are directly relevant to fire matters under the Construction Products Directive:

Guidance paper G -
The European classification system for the reaction to fire performance of construction products.

Guidance paper J -
Transitional arrangements under the Construction Products Directive.

Commission Decision of 8 February 2000 (2000/147/EC) implementing Council Directive 89/106/EEC as regards the classification of the reaction to fire performance of construction products.

Commission Decision of 3 May 2000 (2000/367/EC) implementing Council Directive 89/106/EEC as regards the classification of the resistance to fire performance of construction products, construction works and parts thereof.

Commission Decision of 26 September 2000 (2000/605/EC) amending Decision 96/603/EC establishing the list of products belonging to Classes A "No contribution to fire" provided for in Decision 94/611/EC implementing Article 20 of Council Directive 89/106/EEC on construction products.

Corrigenda – Corrigendum to Commission Decision 2000/147/EC of 8 February 2000 implementing Council Directive 89/106/EEC as regards the classification of the reaction to fire performance of construction products.

The publication and revision of Commission guidance papers and decisions are ongoing and the latest information in this respect can be found by accessing the European Commission's website via the link on the ODPM web site at: www.odpm.gov.uk/bregs/cpd/index.htm

General Introduction

FIRE SAFETY

Arrangement of Sections

0.1 The functional requirements B1 to B5 of Schedule 1 of the Building Regulations are dealt with separately in one or more Sections. The requirement is reproduced at the start of the relevant Sections, followed by an introduction to the subject.

0.2 The provisions set out in this document deal with different aspects of fire safety, with the following aims.

B1: To ensure satisfactory provision of means of giving an alarm of fire and a satisfactory standard of means of escape for persons in the event of fire in a building.

B2: That fire spread over the internal linings of buildings is inhibited.

B3: To ensure the stability of buildings in the event of fire; to ensure that there is a sufficient degree of fire separation within buildings and between adjoining buildings; and to inhibit the unseen spread of fire and smoke in concealed spaces in buildings.

B4: That external walls and roofs have adequate resistance to the spread of fire over the external envelope, and that spread of fire from one building to another is restricted.

B5: To ensure satisfactory access for fire appliances to buildings and the provision of facilities in buildings to assist fire fighters in the saving of life of people in and around buildings.

0.3 Whilst guidance appropriate to each of these aspects is set out separately in this document, many of the provisions are closely interlinked. For example, there is a close link between the provisions for means of escape (B1) and those for the control of fire growth (B2), fire containment (B3), and facilities for the fire service (B5). Similarly there are links between B3 and the provisions for controlling external fire spread (B4), and between B3 and B5. Interaction between these different requirements should be recognised where variations in the standard of provision are being considered. A higher standard under one of the requirements may be of benefit in respect of one or more of the other requirements. The guidance in the document as a whole should be considered as a package aimed at achieving an acceptable standard of fire safety.

0.4 In the guidance on B1 the provisions for dwellings are separated from those for all other types of building, because there are important differences in the approach that has been adopted. Dwellinghouses (Section 2) pose different problems from flats and maisonettes, which are therefore treated separately in Section 3.

Appendices: provisions common to more than one of Part B's requirements

0.5 Guidance on matters that refer to more than one of the Sections is in a series of Appendices, covering the following subjects:

Appendix A. fire performance of materials and structures;

Appendix B. provisions regarding fire doors;

Appendix C. methods of measurement;

Appendix D. a classification of purpose groups;

Appendix E. definitions;

Appendix F. insulating core panels.

Purpose groups

0.6 Much of the guidance in this document is related to the use of the building. The use classifications are termed purpose groups, and they are described in Appendix D.

Fire performance of materials and structures

0.7 Much of the guidance throughout this document is given in terms of performance in relation to standard fire test methods. Details are drawn together in Appendix A to which reference is made where appropriate. In the case of fire protection systems reference is made to standards for systems design and installation. Standards referred to are listed in Appendix G.

Fire doors

0.8 Guidance in respect of fire doors is set out in Appendix B.

Methods of measurement

0.9 Some form of measurement is an integral part of much of the guidance in this document, and methods are set out in Appendix C. Aspects of measurement specific to means of escape are covered in the introduction to B1 (paragraphs B1.xxv onwards).

Definitions

0.10 The definitions are given in Appendix E.

Fire safety engineering

0.11 Fire safety engineering can provide an alternative approach to fire safety. It may be the only practical way to achieve a satisfactory standard of fire safety in some large and complex buildings, and in buildings containing different uses, eg airport terminals. Fire safety engineering may also be suitable for solving a problem with an aspect of the building design which otherwise follows the provisions in this document.

British Standard Draft for Development (DD) 240 *Fire safety engineering in buildings* provides a framework and guidance on the design and assessment of fire safety measures in buildings. Following the discipline of DD 240 should enable designers and building control bodies to be aware of the relevant issues, the need to consider the complete fire-safety system, and to follow a disciplined analytical framework.

0.12 Some variation of the provisions set out in this document may also be appropriate where Part B applies to existing buildings, and particularly in buildings of special architectural or historic interest, where adherence to the guidance in this document might prove unduly restrictive. In such cases it would be appropriate to take into account a range of fire safety features, some of which are dealt with in this document, and some of which are not addressed in any detail, and to set these against an assessment of the hazard and risk peculiar to the particular case.

0.13 Factors that should be taken into account include:

a. the anticipated probability of a fire occurring;

b. the anticipated fire severity;

c. the ability of a structure to resist the spread of fire and smoke;

d. the consequential danger to people in and around the building.

0.14 A wide variety of measures could be considered and incorporated to a greater or lesser extent, as appropriate in the circumstances. These include:

a. the adequacy of means to prevent fire;

b. early fire warning by an automatic detection and warning system;

c. the standard of means of escape;

d. provision of smoke control;

e. control of the rate of growth of a fire;

f. the adequacy of the structure to resist the effects of a fire;

g. the degree of fire containment;

h. fire separation between buildings or parts of buildings;

i. the standard of active measures for fire extinguishment or control;

j. facilities to assist the fire service;

k. availability of powers to require staff training in fire safety and fire routines, eg under the Fire Precautions Act 1971, the Fire Precautions (Workplace) Regulations 1997, or registration or licensing procedures;

l. consideration of the availability of any continuing control under other legislation that could ensure continued maintenance of such systems;

m. management.

0.15 It is possible to use quantitative techniques to evaluate risk and hazard. Some factors in the measures listed above can be given numerical values in some circumstances. The assumptions made when quantitative methods are used need careful assessment.

Shopping complexes and buildings containing one or more atria

0.16 An example of an overall approach to fire safety can be found in BS 5588: Part 10 *Fire precautions in the design, construction and use of buildings. Code of practice for shopping complexes*, which is referred to in Section 12.

Similarly a building containing an atrium passing through compartment floors may need special fire safety measures. Guidance on suitable fire safety measures in these circumstances is to be found in BS 5588: Part 7 *Code of practice for the incorporation of atria in buildings.* (See also paragraph 9.8.)

Hospitals

0.17 The design of fire safety in hospitals is covered by Health Technical Memorandum (HTM) 81 *Fire precautions in new hospitals* (revised 1996) Where the guidance in that document is followed, Part B of the Building Regulations will be satisfied.

Property protection

0.18 Building Regulations are intended to ensure that a reasonable standard of life safety is provided, in case of fire. The protection of property, including the building itself, may require additional measures, and insurers will in general seek their own higher standards, before accepting the insurance risk. Guidance is given in the LPC *Design guide for the fire protection of buildings*.

Guidance for asset protection in the Civil and Defence Estates is given in the *Crown Fire Standards* published by the Property Advisers to the Civil Estate (PACE).

Material alteration

0.19 An alteration which results in a building being less satisfactory in relation to compliance with the requirements of Parts B1, B3, B4 or B5, than it was before, is controllable under Regulations 3 (meaning of building work) and 4 (requirements relating to building work) of the Building Regulations, as a material alteration.

Performance of protection systems, materials and structures

0.20 Since the performance of a system, product, component or structure is dependent upon satisfactory site installation, testing and maintenance, independent schemes of certification and registration of installers and maintenance firms of such will provide confidence in the appropriate standard of workmanship being provided.

Confidence that the required level of performance can be achieved, will be demonstrated by the use of a system, material, product or structure which is provided under the arrangements of a product conformity certification scheme and an accreditation and registration of installers scheme.

Third party accredited product conformity certification schemes not only provide a means of identifying materials and designs of systems, products or structures which have demonstrated that they have the requisite performance in fire, but additionally provide confidence that the systems, materials, products or structures actually supplied are provided to the same specification or design as that tested/assessed.

Third party accreditation and registration of installers of systems, materials, products or structures provide a means of ensuring that installations have been conducted by knowledgeable contractors to appropriate standards, thereby increasing the reliability of the anticipated performance in fire.

The Requirement

This Approved Document deals with the following Requirement from Part B of Schedule 1 to the Building Regulations 2000.

Requirement	Limits on application
Means of warning and escape **B1.** The building shall be designed and constructed so that there are appropriate provisions for the early warning of fire, and appropriate means of escape in case of fire from the building to a place of safety outside the building capable of being safely and effectively used at all material times.	Requirement B1 does not apply to any prison provided under section 33 of the Prisons Act 1952 (power to provide prisons etc.).

Guidance

Performance

In the Secretary of State's view the requirement of B1 will be met if:

a. there are routes of sufficient number and capacity, which are suitably located to enable persons to escape to a place of safety in the event of fire;

b. the routes are sufficiently protected from the effects of fire by enclosure where necessary;

c. the routes are adequately lit;

d. the exits are suitably signed; and if

e. there are appropriate facilities to either limit the ingress of smoke to the escape route(s) or to restrict the fire and remove smoke;

all to an extent necessary that is dependent on the use of the building, its size and height; and

f. there is sufficient means for giving early warning of fire for persons in the building.

Introduction

B1.i These provisions relate to building work and material changes of use which are subject to the functional requirement B1, and they may therefore affect new or existing buildings. They are concerned with the measures necessary to ensure reasonable facilities for means of escape in case of fire. They are only concerned with structural fire precautions where these are necessary to safeguard escape routes.

They assume that in the design of the building, reliance should not be placed on external rescue by the fire service. This Approved Document has been prepared on the basis that, in an emergency, the occupants of any part of a building should be able to escape safely without any external assistance.

Special considerations, however, apply to some institutional buildings in which the principle of evacuation without assistance is not practical.

It should also be noted that the guidance for a typical 1 or 2 storey dwelling is limited to the provision of smoke alarms and to the provision of openable windows for emergency egress.

Interaction with other legislation

B1.ii Attention is drawn to the fact that there may be legislation, other than the Building Regulations, imposing requirements for means of escape in case of fire and other fire safety measures, with which the building must comply, and which will come into force when the building is occupied.

The Fire Precautions Act 1971 and the Fire Precautions (Workplace) Regulations 1997 as amended in 1999, will apply to certain premises (other than dwellings) to which the guidance contained in this document applies. The Fire Authority is responsible for the enforcement of both the above Act and the Regulations.

There are also other Acts and Regulations that impose fire safety requirements as a condition of a licence or registration. Whilst this other legislation is enforced by a number of different authorities, in the majority of cases the applicant and/or enforcing authority is required to consult the Fire Authority before a licence or registration is granted.

B1.iii Under the Fire Precautions Act 1971, the Fire Authority cannot, as a condition for issuing a certificate, make requirements for structural or other alterations to the fire precautions arrangements, if the aspects of the fire precautions concerned have been the subject of a Building Regulation approval. However if the Fire Authority is satisfied that the fire precautions are inadequate by reason of matters that were not subject to a Building Regulation approval or were not known at the time of the approval, then the Fire Authority is not barred from making requirements.

In those premises subject to the Fire Precautions (Workplace) Regulations 1997 the occupier is required to undertake and continually review a risk assessment to ensure that the employees within the premises are not placed at risk from fire. This risk assessment must allow for changes to the fire safety measures provided, subject to the risks identified. In premises subject to these Regulations, as in the case of premises subject to the other legislation containing fire safety requirements, the enforcing authority is not subject to any restriction on the provision of additional fire safety measures.

It should be noted that it is possible for a building to be subject to the Fire Precautions Act, the Fire Precautions (Workplace) Regulations and other legislation imposing fire safety requirements at the same time.

Taking the above into account, it is therefore recommended that the applicant ensures that the fire precautions incorporated into any proposed building works meet the requirements of all those authorities that may be involved in the enforcement of other fire safety related legislation, and that consultation with those authorities takes place in conjunction with the Building Regulation approval.

In addition, a requirement for consultation between enforcing bodies is contained in the Fire Precautions Act 1971 and the Fire Precautions (Workplace) Regulations 1997.

Guidance on the consultation procedures that should be adopted to ensure that the requirements of all enforcing authorities are addressed at Building Regulation Approval stage is contained in *Building Regulation and Fire Safety – Procedural Guidance*, published jointly by the Department of the Environment, Transport and the Regions, the Home Office and the Welsh Office.

B1.iv Under the Health and Safety at Work etc Act 1974 the Health and Safety Executive may have similar responsibilities for certification in the case of highly specialised industrial and storage premises.

B1.v Under the Housing Act 1985 the local authority is obliged to require means of escape in case of fire in certain types of houses which are occupied by persons not forming a single household (Houses in Multiple Occupation).

However, compliance with the guidance in this document will enable a newly constructed or converted House in Multiple Occupation to achieve an acceptable standard of fire safety.

B1.vi There are a number of other Statutes enforced by the local authority or the fire authority that may be applied to premises of specific uses once they are occupied.

Management of premises

B1.vii This Approved Document has been written on the assumption that the building concerned will be properly managed. Failure to take proper management responsibility may result in the prosecution of a building owner or occupier under legislation such as the Fire Precautions Act or the Health and Safety at Work etc Act, and/or prohibition of the use of the premises.

Analysis of the problem

B1.viii The design of means of escape, and the provision of other fire safety measures such as a fire alarm system (where appropriate), should be based on an assessment of the risk to the occupants should a fire occur. The assessment should take into account the nature of the building structure, the use of the building, the processes undertaken and/or materials stored in the building; the potential sources of fire; the potential of fire spread through the building; and the standard of fire safety management proposed. Where it is not possible to identify with any certainty any of these elements a judgement as to the likely level of provision must be made.

B1.ix Fires do not normally start in two different places in a building at the same time. Initially a fire will create a hazard only in the part in which it starts and it is unlikely, at this stage, to involve a large area. The fire may subsequently spread to other parts of the building, usually along the circulation routes. The items that are the first to be ignited are often furnishings and other items not controlled by the regulations. It is less likely that the fire will originate in the structure of the building itself and the risk of it originating accidentally in circulation areas, such as corridors, lobbies or stairways, is limited, provided that the combustible content of such areas is restricted.

B1.x The primary danger associated with fire in its early stages is not flame but the smoke and noxious gases produced by the fire. They cause most of the casualties and may also obscure the way to escape routes and exits. Measures designed to provide safe means of escape must therefore provide appropriate arrangements to limit the rapid spread of smoke and fumes.

Criteria for means of escape

B1.xi The basic principles for the design of means of escape are:

a. that there should be alternative means of escape from most situations;

b. where direct escape to a place of safety is not possible, it should be possible to reach a place of relative safety, such as a protected stairway, which is on a route to an exit, within a reasonable travel distance. In such cases the means of escape will consist of two parts, the first being unprotected in accommodation and circulation areas, and the second in protected stairways (and in some circumstances protected corridors).

The ultimate place of safety is the open air clear of the effects of the fire. However, in modern buildings which are large and complex, reasonable safety may be reached within the building, provided suitable planning and protection measures are incorporated.

B1.xii The following are not acceptable as means of escape:

a. lifts (except for a suitably designed and installed evacuation lift that may be used for the evacuation of disabled people, in a fire);

b. portable ladders and throw-out ladders; and

c. manipulative apparatus and appliances: eg fold down ladders and chutes.

Escalators should not be counted as providing predictable exit capacity, although it is recognised that they are likely to be used by people who are escaping. Mechanised walkways could be accepted, and their capacity assessed on the basis of their use as a walking route, while in the static mode.

Alternative means of escape

B1.xiii There is always the possibility of the path of a single escape route being rendered impassable by fire, smoke or fumes and, ideally, people should be able to turn their backs on a fire wherever it occurs and travel away from it to a final exit or protected escape route leading to a place of safety. However in certain conditions a single direction of escape (a dead end) can be accepted as providing reasonable safety. These conditions depend on the use of the building and its associated fire risk, the size and height of the building, the extent of the dead end, and the numbers of persons accommodated within the dead end.

Unprotected and protected escape routes

B1.xiv The unprotected part of an escape route is that part which a person has to traverse before reaching either the safety of a final exit or the comparative safety of a protected escape route, ie a protected corridor or protected stairway.

Unprotected escape routes should be limited in extent so that people do not have to travel excessive distances while exposed to the immediate danger of fire and smoke.

Even with protected horizontal escape routes the distance to a final exit or protected stairway needs to be limited because the structure does not give protection indefinitely.

B1.xv Protected stairways are designed to provide virtually 'fire sterile' areas which lead to places of safety outside the building. Once inside a protected stairway, a person can be considered to be safe from immediate danger from flame and smoke. They can then proceed to a place of safety at their own pace. To enable this to be done, flames, smoke and gases must be excluded from these escape routes, as far as is reasonably possible, by fire-resisting structures or by an appropriate smoke control system, or by a combination of both these methods. This does not preclude the use of unprotected stairs for day-to-day circulation, but they can only play a very limited role in terms of means of escape due to their vunerability in fire situations.

Means of escape for disabled people

B1.xvi Part M of the Regulations, Access and facilities for disabled people, requires reasonable provision for access by disabled people to certain buildings, or parts of buildings. However it may not be necessary to incorporate special structural measures to aid means of escape for the disabled. Management arrangements to provide assisted escape may be all that is necessary. BS 5588: Part 8 *Fire precautions in the design, construction and use of buildings, Code of practice for means of escape for disabled people*, gives guidance on means of escape for disabled people in all premises other than dwellings. It introduces the concept of refuges and the use of an evacuation lift, and stresses the need for effective management of the evacuation.

Security

B1.xvii The need for easy and rapid evacuation of a building in case of fire may conflict with the control of entry and exit in the interest of security. Measures intended to prevent unauthorised access can also hinder entry of the fire service to rescue people trapped by fire.

Potential conflicts should be identified and resolved at the design stage and not left to ad hoc expedients after completion. The architectural liaison officers attached to most police forces are a valuable source of advice.

It is not appropriate to seek to control the type of lock used on front doors to dwellings under the Building Regulations. Some more detailed guidance on door security in buildings other than single family dwelling houses is given in paragraphs 6.11 and 6.12.

Alternative approaches

B1.xviii The Building Regulations requirements for means of escape will be satisfied by following the relevant guidance given in either the publications in paragraphs B1.xix – B1.xxiii or in Sections 2 – 6 of this Approved Document.

General

B1.xix BS 5588: *Part 0 Fire precautions in the design, construction and use of buildings, Guide to fire safety codes of practice for particular premises/applications* includes reference to various codes and guides dealing with the provision of means of escape. If one of those codes or guides is adopted, the relevant recommendations concerning means of escape in case of fire in the particular publication should be followed, rather than a mixture of the publication and provisions in the relevant sections of this Approved Document. However, there may be circumstances where it is necessary to use one publication to supplement another, as with the use in Section 18 of BS 5588: Part 5 *Code of practice for firefighting stairs and lifts*.

Note: Buildings for some particular industrial and commercial activities presenting a special fire hazard eg those involved with the sale of fuels, may require additional fire precautions to those detailed in this Approved Document. Reference to guidance for such building applications is given in BS 5588: Part 0.

Hospitals

B1.xx In parts of hospitals designed to be used by patients, and in similar accommodation such as nursing homes and homes for the elderly, where there are people who are bedridden or who have very restricted mobility, the principle of total evacuation of a building in the event of fire may be inappropriate. It is also unrealistic to suppose that all patients will leave without assistance.

In this and other ways the specialised nature of some health care premises demands a different approach to the provision of means of escape, from much of that embodied by the guidance in this Approved Document.

NHS Estates has prepared a set of guidance documents on fire precautions in Health Care buildings, under the general title of *"Firecode"*, taking into account the particular characteristics of these buildings. These documents may also be used for non-NHS health care premises.

The provision of means of escape in new hospitals should therefore follow the guidance in Firecode HTM 81 *Fire precautions in new hospitals*. Where work to existing hospitals is concerned with means of escape, the guidance in the appropriate section of the relevant Firecode should be followed. Attention is also drawn to the Home Office *Draft guide to fire precautions in existing residential care premises* which is under review.

Where a house of one or two storeys is converted for use as an unsupervised Group Home for not more than 6 mentally impaired or mentally ill people, it should be regarded as a purpose group 1(c) building if the means of escape are provided in accordance with HTM 88 *Guide to fire precautions in NHS housing in the community for mentally handicapped (or mentally ill) people*. Where the building is new, it may be more appropriate to regard it as being in purpose group 2(b).

Note: Firecode contains managerial and other fire safety provisions which are outside the scope of building regulations.

Shopping complexes

B1.xxi Although the guidance in this Approved Document may be readily applied to individual shops, shopping complexes present a different set of escape problems. A suitable approach is given in section 4 of BS 5588: Part 10: 1991 *Fire precautions in the design, construction and use of buildings, Code of practice for shopping complexes*.

Note: BS 5588: Part 10 applies more restrictive provisions to units with only one exit in covered shopping complexes than given in BS 5588: Part 11 *Code of practice for shops, offices, industrial, storage and other similar buildings*.

Assembly buildings

B1.xxii There are particular problems that arise when people are limited in their ability to escape by fixed seating. This may occur at sports events, theatres, lecture halls and conference centres etc. Guidance on this and other aspects of means of escape in assembly buildings is given in sections 3 and 5 of BS 5588: Part 6: 1991 *Code of practice for places of assembly* and the relevant recommendations concerning means of escape in case of fire of that code should be followed, in appropriate cases. The guidance given in the *Guide to fire precautions in existing places of entertainment and like premises* (HMSO) may also be followed.

In the case of buildings to which the Safety of Sports Grounds Act 1975 applies, the guidance in the *Guide to safety at sports grounds* (HMSO) should also be followed.

Schools and other education buildings

B1.xxiii By following the guidance in this Approved Document it is possible to meet the fire safety objectives of the Department for Education and Employment's constructional standards for schools.

Use of the document

B1.xxiv Section 1 deals with fire alarm and fire detection systems in all buildings. Sections 2 & 3 deal with means of escape from dwellings and Sections 4 & 5 with buildings other than dwellings. Section 2 is about dwellinghouses and Section 3 is on flats and maisonettes. Section 4 concerns the design of means of escape on one level (the horizontal phase in multi-storey buildings). Section 5 deals with stairways and the vertical phase of the escape route. Section 6 gives guidance on matters common to all parts of the means of escape, other than in houses.

Methods of measurement

B1.xxv The following methods of measurement apply specifically to B1. Other aspects of measurement applicable to Part B in general are given in Appendix C.

Occupant capacity

B1.xxvi The **occupant capacity** of a room, storey, building or part of a building is:

a. the maximum number of persons it is designed to hold; or

b. the number calculated by dividing the area of room or storey(s) (m^2) by a floor space factor (m^2 per person) such as those given in Table 1 for guidance.

Note: 'area' excludes stair enclosures, lifts, sanitary accommodation and any other fixed part of the building structure (but counters and display units etc should not be excluded).

Travel distance

B1.xxvii Travel distance is measured by way of the shortest route which if:

a. there is fixed seating or other fixed obstructions, is along the centre line of the seatways and gangways;

b. it includes a stair, is along the pitch line on the centre line of travel.

Width

B1.xxviii The width of:

a. a **door (or doorway)** is the clear width when the door is open (see Diagram 1);

b. an **escape route** is the width at 1500mm above floor level when defined by walls or, elsewhere, the minimum width of passage available between any fixed obstructions;

c. a **stair** is the clear width between the walls or balustrades.

B1 INTRODUCTION

Notes:

1. Door hardware, handrails and strings which do not intrude more than 100mm into these widths may be ignored (see Diagram 1).

2. The rails used for guiding a stair lift may be ignored when considering the width of a stair. However, it is important that the chair or carriage is able to be parked in a position that does not cause an obstruction to either the stair or landing.

Diagram 1 Measurement of door width

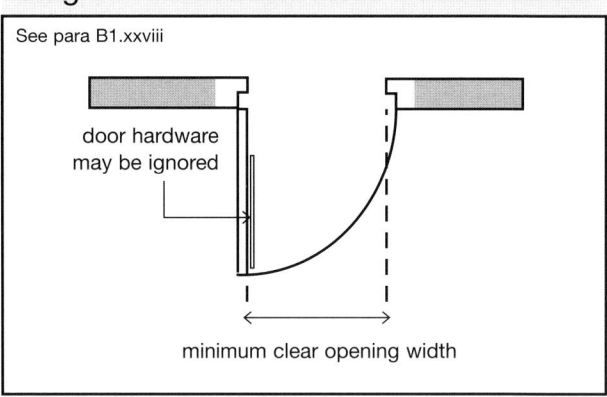

See para B1.xxviii

door hardware may be ignored

minimum clear opening width

Table 1 Floor space factors (1)

	Type of accommodation (2)(3)	Floor space factor m²/person
1.	Standing spectator areas, Bars without seating and similar refreshment areas	0.3
2.	Amusement arcade, Assembly hall (including a general purpose place of assembly), Bingo hall, Club, Crush hall, Dance floor or hall, Venue for pop concert and similar events	0.5
3.	Concourse, Queuing area or Shopping mall (4)	0.7
4.	Committee room, Common room, Conference room, Dining room, Licensed betting office (public area), Lounge or bar (other than in 1 above), Meeting room, Reading room, Restaurant, Staff room or Waiting room (5)	1.0
5.	Exhibition hall or Studio (film, radio, television, recording)	1.5
6.	Skating rink	2.0
7.	Shop sales area (6)	2.0
8.	Art gallery, Dormitory, Factory production area, Museum or Workshop	5.0
9.	Office	6.0
10.	Shop sales area (7)	7.0
11.	Kitchen or Library	7.0
12.	Bedroom or Study-bedroom	8.0
13.	Bed-sitting room, Billiards or snooker room or hall	10.0
14.	Storage and warehousing	30.0
15.	Car park	two persons per parking space

Notes:

1. As an alternative to using the values in the table, the floor space factor may be determined by reference to actual data taken from similar premises. Where appropriate, the data should reflect the average occupant density at a peak trading time of year.

2. Where accommodation is not directly covered by the descriptions given, a reasonable value based on a similar use may be selected.

3. Where any part of the building is to be used for more than one type of accommodation, the most onerous factor(s) should be applied. Where the building contains different types of accommodation, the occupancy of each different area should be calculated using the relevant space factor.

4. Refer to section 4 of BS 5588: Part 10: 1991 *Code of practice for shopping complexes* for detailed guidance on the calculation of occupancy in common public areas in shopping complexes.

5. Alternatively the occupant capacity may be taken as the number of fixed seats provided, if the occupants will normally be seated.

6. Shops excluding those under item 10, but including – supermarkets and department stores (main sales areas), shops for personal services such as hairdressing and shops for the delivery or collection of goods for cleaning, repair or other treatment or for members of the public themselves carrying out such cleaning, repair or other treatment.

7. Shops (excluding those in covered shopping complexes but including department stores) trading predominantly in furniture, floor coverings, cycles, prams, large domestic appliances or other bulky goods, or trading on a wholesale self-selection basis (cash and carry).

Section 1

FIRE ALARM AND FIRE DETECTION SYSTEMS

Introduction

1.1 Provisions are made in this section for suitable arrangements to be made in all buildings to give early warning in the event of fire.

Paragraphs 1.2 to 1.22 deal with dwellings and paragraphs 1.23 to 1.32 with other buildings.

Dwellings

General

1.2 In most houses the installation of smoke alarms or automatic fire detection and alarm systems, can significantly increase the level of safety by automatically giving an early warning of fire.

1.3 If houses are not protected by an automatic fire detection and alarm system in accordance with the relevant recommendations of BS 5839: Part 1 *Fire detection and alarm systems for buildings, Code of practice for system design, installation and servicing* to at least an L3 standard, or BS 5839: Part 6 *Code of practice for the design and installation of fire detection and alarm systems in dwellings* to at least a Grade E type LD3 standard, they should be provided with a suitable number of smoke alarms installed in accordance with the guidance in paragraphs 1.4 to 1.22 below.

1.4 The smoke alarms should be mains-operated and conform to BS 5446 *Components of automatic fire alarm systems for residential premises*, Part 1 *Specification for self-contained smoke alarms and point-type smoke detectors*. They may have a secondary power supply such as a battery (either rechargeable or replaceable) or capacitor. More information on power supplies is given in clause 13 of BS 5839: Part 6: 1995.

Note: BS 5446: Part 1 covers smoke alarms based on ionization chamber smoke detectors and optical (photo-electric) smoke detectors. The different types of detector respond differently to smouldering and fast flaming fires. Either type of detector is generally suitable. However, the choice of detector type should, if possible, take into account the type of fire that might be expected and the need to avoid false alarms. Optical detectors tend to be less affected by low levels of 'invisible' smoke that often cause false alarms.

BS 5839: Part 6 suggests that, in general, optical smoke alarms should be installed in circulation spaces such as hallways and landings, and ionization chamber based smoke alarms may be the more appropriate type in rooms, such as the living room or dining room where a fast burning fire may present a greater danger to occupants than a smouldering fire.

Large houses

1.5 A house may be regarded as large if any of its storeys exceed 200m².

1.6 A large house of more than 3 storeys (including basement storeys) should be fitted with an L2 system as described in BS 5839: Part 1: 1988, except that the provisions in clause 16.5 regarding duration of the standby supply need not be followed. However with unsupervised systems, the standby supply should be capable of automatically maintaining the system in normal operation (though with audible and visible indication of failure of the mains) for 72 hours, at the end of which sufficient capacity remains to supply the maximum alarm load for at least 15 minutes.

1.7 A large house of no more than 3 storeys (including basement storeys) may be fitted with an automatic fire detection and alarm system of Grade B type LD3 as described in BS 5839: Part 6 instead of an L2 system.

Loft conversions

1.8 Where a loft in a one or two storey house is converted into habitable accommodation, an automatic smoke detection and alarm system based on linked smoke alarms should be installed (see paragraph 2.26).

Flats and maisonettes

1.9 The same principles apply within flats and maisonettes as for houses, while noting that:

a. the provisions are not intended to be applied to the common parts of blocks of flats and do not include interconnection between installations in separate flats;

b. a flat with accommodation on more than one level (ie a maisonette) should be treated in the same way as a house with more than one storey.

Note: Some student residential accommodation is constructed in the same way as a block of flats. Where groups of students share one flat with its own entrance door, it is appropriate to provide an automatic detection system within each flat. In student flats constructed on the compartmentation principles for flats in Section 9 (B3), the automatic detection system will satisfy the requirements of building regulations if it gives a warning in the flat of fire origin. Where a general evacuation is required, the alarm system should follow the guidance in paragraph 1.30.

Sheltered housing

1.10 The detection equipment in a sheltered housing scheme with a warden or supervisor, should have a connection to a central monitoring point (or central alarm relay station) so that the person in charge is aware that a fire has been detected in one of the dwellings, and can identify the dwelling concerned. These provisions are not intended to be applied to the common parts of a sheltered housing development, such as communal lounges, or to sheltered accommodation in the Institutional or Other residential purpose groups.

Installations based on smoke alarms

1.11 Smoke alarms should normally be positioned in the circulation spaces between sleeping spaces and places where fires are most likely to start (eg kitchens and living rooms) to pick up smoke in the early stages, while also being close enough to bedroom doors for the alarm to be effective when occupants are asleep.

1.12 In a house (including bungalows) there should be at least one smoke alarm on every storey.

1.13 Where more than one smoke alarm is installed they should be linked so that the detection of smoke by one unit operates the alarm signal in all of them. The manufacturers' instructions about the maximum number of units that can be linked should be observed.

1.14 Smoke alarms should be sited so that:

a. there is a smoke alarm in the circulation space within 7.5m of the door to every habitable room;

b. where the kitchen area is not separated from the stairway or circulation space by a door, there should be a compatible interlinked heat detector in the kitchen, in addition to whatever smoke alarms are needed in the circulation space(s);

c. they are ceiling mounted and at least 300mm from walls and light fittings (unless in the case of light fittings there is test evidence to prove that the proximity of the light fitting will not adversely affect the efficiency of the detector). Units designed for wall mounting may also be used provided that the units are above the level of doorways opening into the space, and they are fixed in accordance with manufacturers' instructions; and

d. the sensor in ceiling mounted devices is between 25mm and 600mm below the ceiling (25-150mm in the case of heat detectors).

    Note: This guidance applies to ceilings that are predominantly flat and horizontal.

1.15 It should be possible to reach the smoke alarms to carry out routine maintenance, such as testing and cleaning, easily and safely. For this reason smoke alarms should not be fixed over a stair shaft or any other opening between floors.

1.16 Smoke alarms should not be fixed next to or directly above heaters or air conditioning outlets. They should not be fixed in bathrooms, showers, cooking areas or garages, or any other place where steam, condensation or fumes could give false alarms.

Smoke alarms should not be fitted in places that get very hot (such as a boiler room), or very cold (such as an unheated porch). They should not be

fixed to surfaces which are normally much warmer or colder than the rest of the space, because the temperature difference might create air currents which move smoke away from the unit.

A requirement for maintenance can not be made as a condition of passing plans by the Building Control Body. However the attention of developers and builders is drawn to the importance of providing the occupants with information on the use of the equipment, and on its maintenance (or guidance on suitable maintenance contractors).

Note: BS 5839: Part 1 and Part 6 recommend that occupiers should receive the manufacturers' instructions concerning the operation and maintenance of the alarm system.

Power supplies

1.17 The power supply for a smoke alarm system should be derived from the dwelling's mains electricity supply. The mains supply to the smoke alarm(s) should comprise a single independent circuit at the dwelling's main distribution board (consumer unit). If the smoke alarm installation does not include a stand-by power supply, no other electrical equipment should be connected to this circuit (apart from a dedicated monitoring device installed to indicate failure of the mains supply to the smoke alarms – see below).

1.18 A smoke alarm, or smoke alarm system, that includes a standby power supply or supplies, can operate during mains failure. It can therefore be connected to a regularly-used local lighting circuit. This has the advantage that the circuit is unlikely to be disconnected for any prolonged period.

1.19 Devices for monitoring the mains supply to the smoke alarm system may comprise audible or visible signals on each unit or on a dedicated mains monitor connected to the smoke alarm circuit. The circuit design of any mains failure monitor should avoid any significant reduction in the reliability of the supply, and should be sited so that the warning of failure is readily apparent to the occupants. If a continuous audible warning is given, it should be possible to silence it.

1.20 The smoke alarm circuit should preferably not be protected by any residual current device (rcd). However if electrical safety requires the use of a rcd, either:

a. the smoke alarm circuit should be protected by a single rcd which serves no other circuit; or

b. the rcd protection of a smoke alarm circuit should operate independently of any rcd protection for circuits supplying socket-outlets or portable equipment.

1.21 Any cable suitable for domestic wiring may be used for the power supply and interconnection to smoke alarm systems. It does not need any particular fire survival properties. Any conductors used for interconnecting alarms (signalling) should be readily distinguishable from those supplying mains power, eg by colour coding.

Note: Smoke alarms may be interconnected using radio-links, provided that this does not reduce the lifetime or duration of any standby power supply.

1.22 Other effective, though possibly more expensive, options exist. For example, the mains supply may be reduced to extra low voltage in a control unit incorporating a standby trickle-charged battery, before being distributed at that voltage to the alarms.

Buildings other than dwellings

General

1.23 To select the appropriate type of fire alarm/detection system that should be installed into a particular building, the type of occupancy and means of escape strategy (eg simultaneous, phased or progressive horizontal evacuation) must be determined.

For example, if occupants normally sleep on the premises eg residential accommodation, the threat posed by a fire is much greater than that in premises where the occupants are normally alert. Where the means of escape is based on simultaneous evacuation, then operation of a manual call point or fire detector should give an almost instantaneous warning from all the fire alarm sounders. However, where the means of escape is based on phased evacuation, then a staged alarm system is appropriate. Such a system enables two or more stages of alarm to be given within a particular area, eg "alert" or "evacuate" signals.

Note: the term fire detection system is used here to describe any type of automatic sensor network and associated control and indicating equipment. Sensors may be sensitive to smoke, heat or radiation. Normally the control and indicating equipment operates a fire alarm system, and it may perform other signalling or control functions as well. Automatic sprinkler systems can also be used to operate a fire alarm system.

1.24 The factors which have to be considered when assessing what standard of fire alarm or automatic fire detection system is to be provided will vary widely from one set of premises to another. Therefore the appropriate standard will need to be considered on a case by case basis.

Note: Where buildings will be controlled under other legislation when occupied, for example premises which are designated under the Fire Precautions Act 1971 and/or are classified as a workplace under the Fire Precautions (Workplace) Regulations 1997, compatibility is essential between what is provided under the Building Regulations at construction stage, and what is needed upon the completion and occupation of the premises. **This should be achieved by means of early consultation with all relevant interested parties, not least the Building Control Body and the Fire Authority.**

Fire alarm systems

1.25 All buildings should have arrangements for detecting fire. In most buildings fires are detected by people, either through observation or smell, and therefore often nothing more will be needed.

1.26 In small buildings/premises the means of raising the alarm may be simple. For instance, where all occupants are near to each other a shouted warning "FIRE" by the person discovering the fire may be all that is needed. In assessing the situation, it must be determined that the warning can be heard and understood throughout the premises, including for example the toilet areas. In other circumstances, manually operated sounders (such as rotary gongs or handbells) may be used. Alternatively a simple manual call point combined with a bell, battery and charger may be suitable.

In all other cases, the building should be provided with a suitable electrically operated fire warning system with manual call points sited adjacent to exit doors and sufficient sounders to be clearly audible throughout the building.

1.27 An electrically operated fire alarm system should comply with BS 5839: Part 1 *Fire detection and alarm systems for buildings, Code of practice for system design, installation and servicing.*

Call points for electrical alarm systems should comply with BS 5839: Part 2 *Specification for manual call points*, or Type A of BS EN 54 *Fire detection and fire alarm systems – Part 11: Manual call points*, and these should be installed in accordance with BS 5839: Part 1. Type B call points should only be used with the approval of the Building Control Body.

Note 1: BS 5839: Part 1 specifies four types of system, ie type L for the protection of life; type M manual alarm systems; type P for property protection; and type X for multi-occupancy buildings. Type L systems are subdivided into L1 – systems installed throughout the protected building; L2 – systems installed only in defined parts of the protected building (a type L2 system should normally include the coverage required of a type L3 system) and L3 – systems installed only for the protection of escape routes. Type P systems are subdivided into P1 – systems installed throughout the protected building and P2 – systems installed only in defined parts of the protected building.

Note 2: BS EN 54-11 covers two types of call points, Type A (direct operation) in which the change to the alarm condition is automatic (i.e. without the need for further manual action) when the frangible element is broken or displaced; and Type B (indirect operation) in which the change to the alarm condition requires a separate manual operation of the operating element by the user after the frangible element is broken or displaced.

1.28 If it is considered that people might not respond quickly to a fire warning, or where people are unfamiliar with the fire warning arrangements, consideration may be given to installing a voice alarm system. Such a system could form part of a public address system and give both an audible signal and verbal instructions in the event of fire.

The fire warning signal should be distinct from other signals which may be in general use and be accompanied by clear verbal instructions.

If a voice alarm system is to be installed, it should comply with BS 5839: Part 8 *Code of practice for the design, installation and servicing of voice alarm systems*.

1.29 In certain premises, eg large shops and places of assembly, an initial general alarm may be undesirable because of the number of members of the public present. The need for fully trained staff to effect pre-planned procedures for safe evacuation will therefore be essential. Actuation of the fire alarm system will cause staff to be alerted, eg by discreet sounders, personal paging systems etc. Provision will normally be made for full evacuation of the premises by sounders or a message broadcast over the public address system. In all other respects, any staff alarm system should comply with BS 5839: Part 1.

Automatic fire detection and fire alarm systems

1.30 Automatic fire detection and alarms in accordance with BS 5839: Part 1 should be provided in Institutional and Other residential occupancies.

1.31 Automatic fire detection systems are not normally needed in Office, Shop and commercial, Assembly and recreation, Industrial, and Storage and other non-residential occupancies. However, there are often circumstances where a fire detection system in accordance with BS 5839: Part 1 may be needed. For example:

a. to compensate for some departure from the guidance elsewhere in this document;

b. as part of the operating system for some fire protection systems, such as pressure differential systems or automatic door releases;

c. where a fire could break out in an unoccupied part of the premises (eg a storage area or basement that is not visited on a regular basis, or a part of the building that has been temporarily vacated) and prejudice the means of escape from any occupied part(s) of the premises.

Notes:

1. General guidance on the standard of automatic fire detection that **may** need to be provided within a building can be found in the Home Office guides that support the Fire Precautions Act 1971 and the Fire Precautions (Workplace) Regulations 1997 and, in the case of the Institutional purpose group, in "Firecode".

2. Guidance on the provision of automatic fire detection within a building which is designed for phased evacuation can be found in paragraph 5.20.

3. Where an atrium building is designed in accordance BS 5588: *Fire precautions in the design, construction and use of buildings, Part 7 Code of practice for the incorporation of atria in buildings*, then the relevant recommendations in that code for the installation of fire alarm/fire detection systems for the design option(s) selected should be followed.

Design and installation of systems

1.32 It is essential that fire detection and fire warning systems are properly designed, installed and maintained. Where a fire alarm system is installed, an installation and commissioning certificate should be provided. Third party certification schemes for fire protection products and related services are an effective means of providing the fullest possible assurances, offering a level of quality, reliability and safety (see paragraph 0.20).

Section 2

DWELLINGHOUSES

Introduction

2.1 The means of escape from a typical one or two storey house are relatively simple to provide. Few provisions are specified in this document beyond ensuring that means are provided for giving early warning in the event of fire (see Section 1) and that suitable means are provided for emergency egress from each storey.

With increasing height more complex provisions are needed because emergency egress through upper windows becomes increasingly hazardous. It is then necessary to protect the internal stairway. If there are floors more than 7.5m above ground level, the risk that the stairway will become impassable before occupants of the upper parts of the house have escaped is appreciable, and an alternative route from those parts is called for.

2.2 In providing fire protection of any kind in houses it should be recognised that measures which significantly interfere with the day-to-day convenience of the occupants may be less reliable in the long term.

2.3 This guidance is also applicable to the design and construction of houses which are considered to be 'houses in multiple occupation' (HMOs) providing there are no more than 6 residents. For HMOs with a greater number of residents, then additional precautions may be necessary. A house in multiple occupation is defined in section 345 of the Housing Act 1985 as "a house which is occupied by persons who do not form a single household". Guidance on the interpretation of this definition is given in DOE Circular 12/93 *Houses in multiple occupation. Guidance to local housing authorities on managing the stock in their area* and Welsh Office Circular 55/93 *Houses in multiple occupation. Guidance on management strategies*.

Technical guidance for HMOs is given in DOE Circular 12/92 *Houses in multiple occupation. Guidance to local housing authorities on standards of fitness under section 352 of the Housing Act 1985* and Welsh Office Circular 25/92 *Local Government and Housing Act 1989. Houses in multiple occupation: standards of fitness*.

Depending on the nature of the occupants and their management, it may be acceptable to treat an unsupervised group home with up to 6 residents as an ordinary dwelling. However because such places have to be registered, the registration authority should be consulted to establish whether there are any additional fire safety measures that the authority will require.

General provisions

Inner rooms

2.4 A room whose only escape route is through another room is termed an inner room and is at risk if a fire starts in that other room (access room). This situation may arise with open plan layouts and sleeping galleries.

Such an arrangement is only acceptable where the inner room is:

a. a kitchen;

b. a laundry or utility room;

c. a dressing room;

d. a bathroom, wc, or shower room;

e. any other room on a floor not more than 4.5m above ground level which complies with paragraph 2.7, 2.8b or 2.10 as appropriate;or

f. a sleeping gallery which complies with paragraph 2.9.

Balconies and flat roofs

2.5 A flat roof forming part of a means of escape should comply with the following provisions:

a. the roof should be part of the same building from which escape is being made;

b. the route across the roof should lead to a storey exit or external escape route; and

c. the part of the roof forming the escape route and its supporting structure, together with any opening within 3m of the escape route, should provide 30 minutes fire resistance (see Appendix A Table A1).

2.6 Where a balcony or flat roof is provided for escape purposes guarding may be needed, in which case it should meet the provisions in Approved Document K *Protection from falling, collision and impact*.

Provisions for escape from floors not more than 4.5m above ground level

Note: Ground level is explained in Appendix C, Diagram C5.

2.7 Except for kitchens, all habitable rooms in the upper storey(s) of a house served by only one stair should be provided with a window (or external door) which complies with paragraph 2.11.

Note: A single window can be accepted to serve two rooms provided both rooms have their own access to the stairs. A communicating door between the rooms must be provided so that it is possible to gain access to the window without passing through the stair enclosure.

2.8 Except for kitchens, all habitable rooms in the ground storey should either:

a. open directly onto a hall leading to the entrance or other suitable exit; or

b. be provided with a window (or door) which complies with paragraph 2.11.

2.9 Where a sleeping gallery is provided:

a. the gallery should be not more than 4.5m above ground level;

b. the distance between the foot of the access stair to the gallery and the door to the room containing the gallery should not exceed 3m;

c. an alternative exit, or an emergency egress window which complies with paragraph 2.11, is needed if the distance from the head of the access stair to any point on the gallery exceeds 7.5m; and

d. any cooking facilities within a room containing a gallery should either:

   i. be enclosed with fire-resisting construction; or

   ii. be remote from the stair to the gallery and positioned such that they do not prejudice the escape from the gallery.

Basements

2.10 Because of the risk that a single stairway may be blocked by smoke from a fire in the basement or ground storey, if the basement storey contains any habitable room, either provide:

a. an external door or window suitable for egress from the basement (see paragraph 2.11); or

b. a protected stairway leading from the basement to a final exit.

Emergency egress windows and external doors

2.11 Any window provided for emergency egress purposes and any external door provided for escape should comply with the following conditions.

a. The window should have an unobstructed openable area that is at least 0.33m² and at least 450mm high and 450mm wide (the route through the window may be at an angle rather than straight through).The bottom of the openable area should be not more than 1100mm above the floor.

   Note: Approved Document K *Protection from falling, collision and impact* specifies a minimum guarding height of 800mm, except in the case of a window in a roof where the bottom of the opening may be 600mm above the floor.

b. The window or door should enable the person escaping to reach a place free from danger from fire. This is a matter for judgement in each case, but in general a courtyard or back garden from which there is no exit other than through other buildings would have to be at least as deep as the dwelling is high to be acceptable, see Diagram 2.

Diagram 2 **Ground or basement storey exit into an enclosed space**

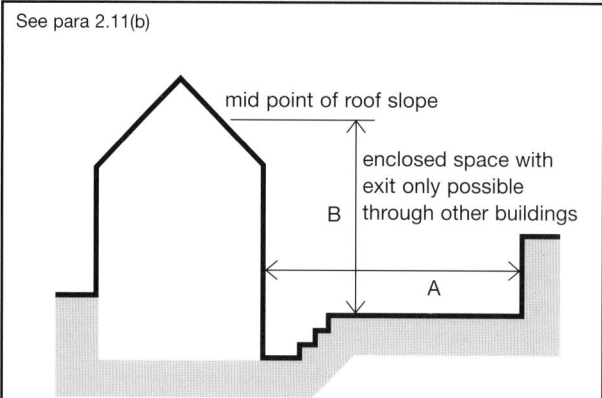

See para 2.11(b)

mid point of roof slope

enclosed space with exit only possible through other buildings

B

A

Depth of back garden A should exceed the height B of the house above ground level for it to be acceptable for an escape route from the ground or basement storey to open into the garden.

---

c. Where provided in a loft conversion of a 2 storey house, a dormer window or roof window should be positioned in accordance with Diagram 6.

Additional provisions for houses with a floor more than 4.5m above ground level

2.12 The provisions described in 2.13-2.16 and 2.17 do not apply if the house has more than one internal stairway which afford effective alternative means of escape and are adequately separated from each other.

Houses with one floor more than 4.5m above ground level

2.13 The house may either have a protected stairway as described in (a) below, or the top floor can be separated and given its own alternative escape route as described in (b). A variation of (b) can be used where the roofspace of an existing two storey dwelling house is being converted into habitable accommodation to form a three storey dwelling house, see paragraphs 2.17 to 2.26.

a. The upper storeys (those above ground storey) should be served by a protected stairway which should either:

    i.   extend to a final exit, see Diagram 3(a), or

    ii. give access to at least two escape routes at ground level, each delivering to final exits and separated from each other by fire-resisting construction and self-closing fire doors, see Diagram 3(b).

b. The top storey should be separated from the lower storeys by fire-resisting construction and be provided with an alternative escape route leading to its own final exit.

    Note: Fire doors in dwellings may be fitted with rising butt hinges rather than spring or other forms of self-closing device (see Appendix E definition of "automatic self-closing device").

Diagram 3 **Alternative arrangements for final exits**

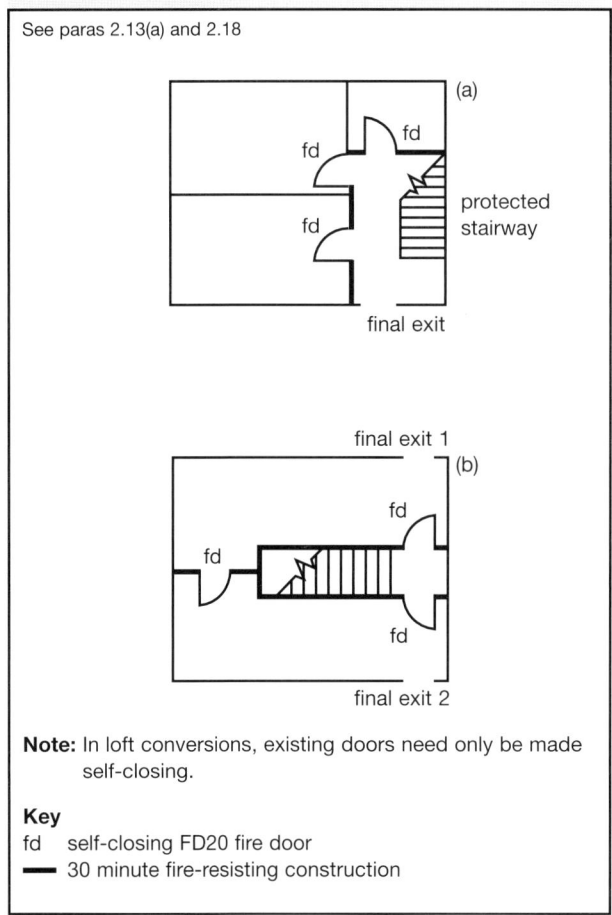

See paras 2.13(a) and 2.18

Note: In loft conversions, existing doors need only be made self-closing.

Key
fd   self-closing FD20 fire door
━   30 minute fire-resisting construction

Houses with more than one floor over 4.5m above ground level

2.14 Where a house has two or more storeys with floors more than 4.5m above ground level (typically a house of four or more storeys), then in addition to meeting the provisions in paragraph 2.13, an alternative escape route should be provided from each storey or level situated 7.5m or more above ground level.

Where the access to the alternative escape route is via:

a. the protected stairway to an upper storey; or

b. a landing within the protected stairway enclosure to an alternative escape route on the same storey; then

c. the protected stairway at or about 7.5m above ground level should be separated from the lower storeys or levels by fire-resisting construction, see Diagram 4.

Diagram 4 **Fire separation in houses with more than one floor over 4.5m above ground level**

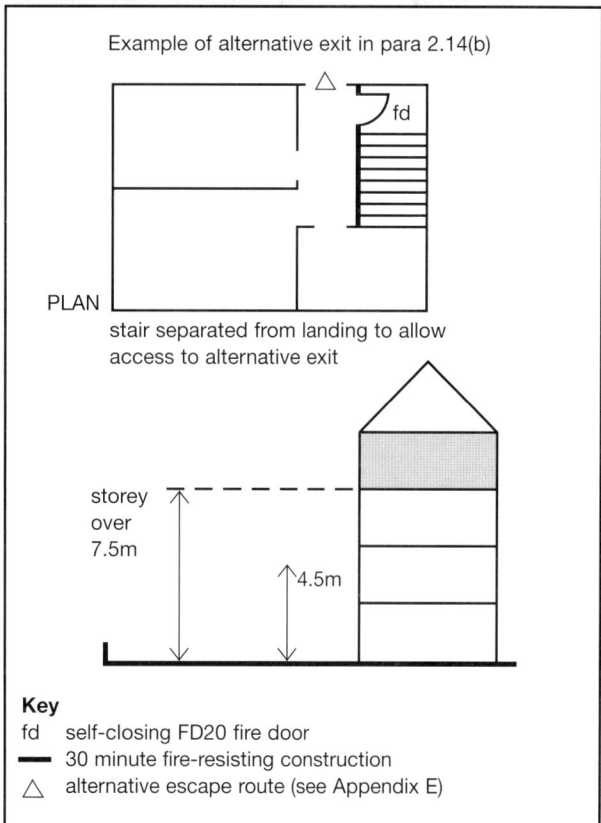

Key
fd   self-closing FD20 fire door
━   30 minute fire-resisting construction
△   alternative escape route (see Appendix E)

Air circulation systems for heating, energy conservation or condensation control in houses with a floor more than 4.5m above ground level

2.15 With these types of systems, the following precautions are needed to avoid the possibility of the system allowing smoke or fire to spread into a protected stairway.

a. Transfer grilles should not be fitted in any wall, door, floor or ceiling enclosing a protected stairway.

b. All ductwork passing through the enclosure to a protected stairway should be so fitted that all joints between the ductwork and the enclosure are fire-stopped.

c. Where ductwork is used to convey air into a protected stairway through the enclosure of the protected stairway, the return air from the protected stairway should be ducted back to the plant.

d. Air and return air grilles or registers should be positioned at a height not exceeding 450mm above floor level.

e. A room thermostat for a ducted warm air heating system should be mounted in the living room at a height between 1370mm and 1830mm, and its maximum setting should not exceed 27°C.

f. Any system of mechanical ventilation which recirculates air should comply with the relevant recommendations in BS 5588 *Fire precautions in the design, construction and use of buildings*, Part 9 *Code of practice for ventilation and air conditioning ductwork.*

Passenger lifts

2.16 Where a passenger lift is provided in the house and it serves any floor more than 4.5m above ground level, it should either be located in the enclosure to the protected stairway (see paragraph 2.13) or be contained in a fire-resisting lift shaft.

Loft conversions

2.17 In the case of an existing two storey house to which a storey is to be added by converting the existing roof space into habitable rooms, the following provisions 2.18-2.25 can be applied as an alternative to those in paragraph 2.13.

However, these alternative provisions are not suitable if:

a. the new second storey exceeds 50m^2 in floor area; or

b. the new second storey is to contain more than two habitable rooms.

Enclosure of existing stair

2.18 The stair in the ground and first storeys should be enclosed with walls and/or partitions which are fire-resisting, and the enclosure should either:

a. extend to a final exit, see Diagram 3(a); or

b. give access to at least two escape routes at ground level, each delivering to final exits and separated from each other by fire-resisting construction and self-closing fire doors, see Diagram 3(b).

Doorways

2.19 Every doorway within the enclosure to the existing stair should be fitted with a door which, in the case of doors to habitable rooms, should be fitted with a self-closing device.

Note: Rising butt hinges are adequate as self-closing devices (see Note to paragraph 2.13).

Any new door to a habitable room should be a fire door. Existing doors need only be fitted with self-closing devices. Existing glazed doors may need to have the glazing changed, see paragraph 2.20.

Glazing

2.20 Any glazing (whether new or exisiting) in the enclosure to the existing stair, including all doors (whether or not they need to be fire doors), but excluding glazing to a bathroom or wc, should be fire-resisting and retained by a suitable glazing system and beads compatible with the type of glass. (See also Appendix A, Table A4.)

New stair

2.21 The new storey should be served by a stair (which may be an alternating tread stair or fixed ladder) meeting the provisions in Approved Document K, *Protection from falling, collision and impact*. The new stair may be located either in a continuation of the existing stairway, or in an enclosure that is separated from the existing stairway, and from ground and first floor accommodation, but which opens into the existing stairway at first floor level, see Diagram 5.

Diagram 5 **Alternatives for the fire separation of the stair and new storey in house conversion**

See paras 2.21 and 2.22

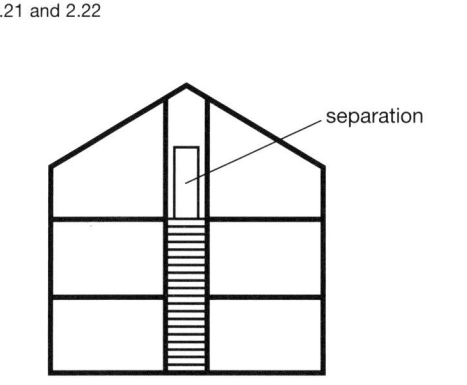

a. stair to new storey is within existing stairway. Separation of accommodation on new storey from stair is at top level.

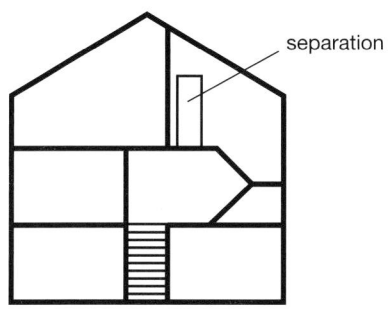

b. stair to new storey is in space converted and separated from first floor accommodation. New storey separated from stair at top level.

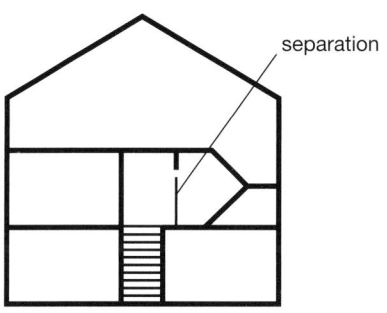

c. stair to new storey is in space converted and separated from first floor accommodation. New storey open to new stair. New stair separated from original stair at first floor level.

Fire separation of new storey

2.22 The new storey should be separated from the rest of the house by fire-resisting construction, see paragraph 8.7. To maintain this separation, measures should be taken to prevent smoke and fire in the stairway from entering the new storey. This may be achieved by providing a self-closing fire door set in fire-resisting construction either at the top or the bottom of the new stair, depending on the layout of the new stairway, see Diagram 5.

Emergency egress windows

2.23 Windows provided for emergency egress purposes from basement, ground or first storeys, provide a means of self-rescue. At higher level escape may depend on a ladder being set up. While this is a departure from the general principle that escape should be without outside assistance it is considered that, in the case of a three storey domestic residential loft conversion this is reasonable as an emergency measure. A fixed ladder on the slope of the roof is not recommended.

2.24 The room (or rooms) in the new storey should each have an openable window or rooflight which meets the relevant provisions in Diagram 6. A door to a roof terrace is also acceptable.

In a 2 room loft conversion, a single window can be accepted provided both rooms have their own access to the stairs. A communicating door between the rooms must be provided so that it is possible to gain access to the window without passing through the stair enclosure.

2.25 The window should be located to allow access for rescue by ladder from the ground (there should therefore be suitable pedestrian access to the point at which a ladder would be set, for fire service personnel to carry a ladder from their vehicle, although it should not be assumed that only the fire service will make a rescue).

Escape across the roof of a ground storey extension is acceptable providing the roof is fire-resisting (see paragraph 2.5 and Appendix A, Table A1). The effect of an extension on the ability to escape from windows in other parts of the house (especially from a loft conversion) should be considered.

Automatic smoke detection and alarms

2.26 Smoke alarms should be fitted as described in Section 1.

Diagram 6 Position of dormer window or rooflight suitable for emergency egress purposes from a loft conversion of a 2-storey dwellinghouse

See paras 2.11(c) and 2.24

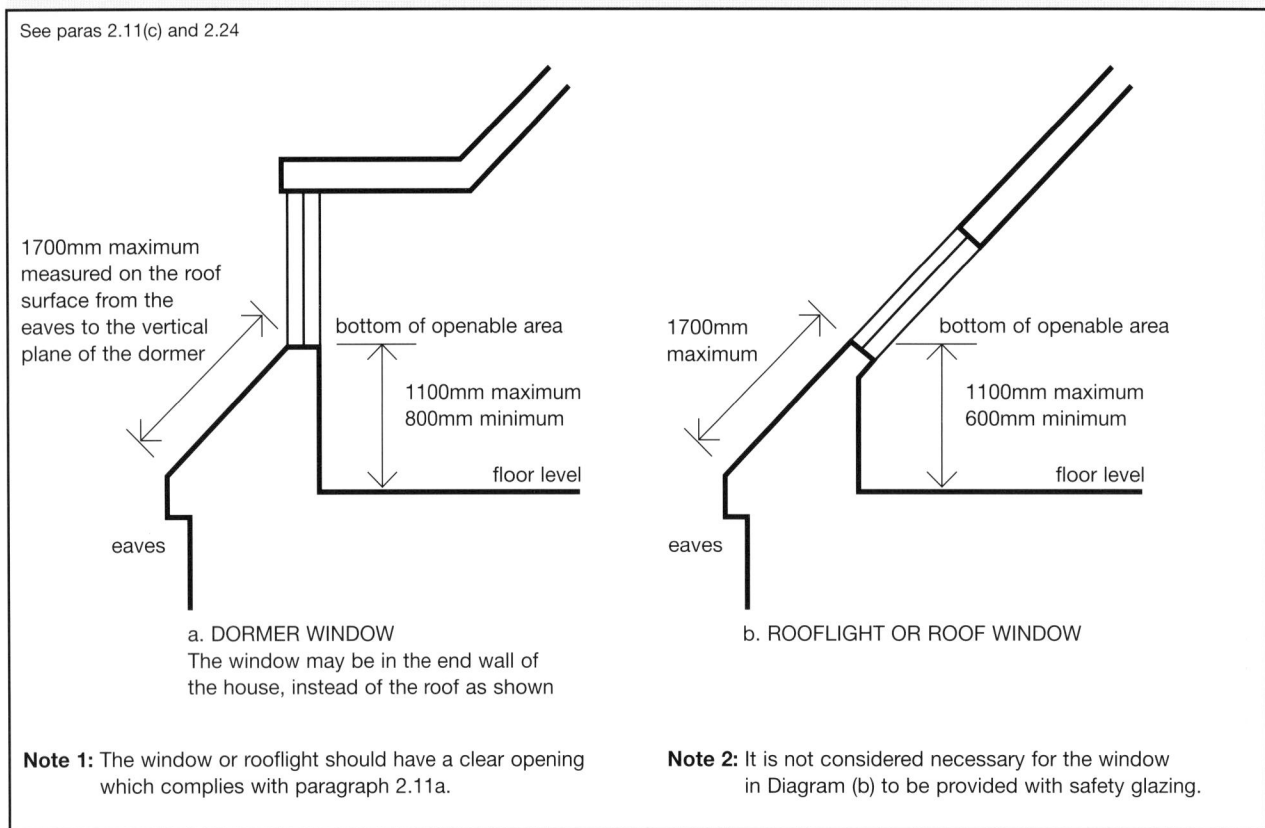

1700mm maximum measured on the roof surface from the eaves to the vertical plane of the dormer

bottom of openable area

1100mm maximum
800mm minimum

floor level

eaves

a. DORMER WINDOW
The window may be in the end wall of the house, instead of the roof as shown

1700mm maximum

bottom of openable area

1100mm maximum
600mm minimum

floor level

eaves

b. ROOFLIGHT OR ROOF WINDOW

Note 1: The window or rooflight should have a clear opening which complies with paragraph 2.11a.

Note 2: It is not considered necessary for the window in Diagram (b) to be provided with safety glazing.

Section 3

FLATS AND MAISONETTES

Introduction

3.1 The means of escape from a flat or maisonette with a floor not more than 4.5m above ground level are relatively simple to provide. Few provisions are specified in this document beyond ensuring that means are provided for giving early warning in the event of fire (see Section 1) and that suitable means are provided for emergency egress from these storeys.

With increasing height more complex provisions are needed because emergency egress through upper windows becomes increasingly hazardous, and in maisonettes internal stairs with a higher level of protection are needed.

3.2 The guidance in this section deals with some common arrangements of flat and maisonette design. Other, less common, arrangements (for example flats entered above or below accommodation level, or flats containing galleries) are acceptable. Guidance on these is given in clauses 9 and 10 of BS 5588 *Fire precautions in the design, construction and use of buildings*, Part 1: 1990 *Code of practice for residential buildings*.

3.3 The provisions for means of escape for flats and maisonettes are based on the assumption that:

a. the fire is generally in a dwelling;

b. there is no reliance on external rescue (eg by a portable ladder);

c. measures in Section 9 (B3) provide a high degree of compartmentation and therefore a low probability of fire spread beyond the dwelling of origin, so that simultaneous evacuation of the building is unlikely to be necessary; and

d. although fires may occur in the common parts of the building, the materials and construction used there should prevent the fabric from being involved beyond the immediate vicinity (although in some cases communal facilities exist which require additional measures to be taken).

3.4 There are two distinct components to planning means of escape from buildings containing flats and maisonettes; escape from within each dwelling, and escape from each dwelling to the final exit from the building.

Paragraphs 3.7 to 3.16 deal with the means of escape within each unit, ie within the private domestic area. Paragraphs 3.17 to 3.48 deal with the means of escape in the common areas of the building.

Houses in multiple occupation

3.5 This guidance is also applicable to flats and maisonettes when they are considered to be houses in multiple occupation. Whether or not a building is a house in multiple occupation depends on the nature of the occupancy, rather than its physical form. See paragraph 2.3 for guidance available for houses in multiple occupation.

Sheltered housing

3.6 Whilst many of the provisions in this Approved Document for means of escape from flats are applicable to sheltered housing, the nature of the occupancy may necessitate some additional fire protection measures. The extent will depend on the form of the development. For example a group of specially adapted bungalows or 2 storey flats, with few communal facilities, need not be treated differently from other 1 or 2 storey houses or flats. Where additional provisions are needed guidance on means of escape can be found in clause 17 of BS 5588: Part 1: 1990.

General provisions

Inner rooms

3.7 A room whose only escape route is through another room is at risk if a fire starts in that other room. The guidance in Section 2, paragraph 2.4, on inner rooms in dwelling houses, applies equally to flats and maisonettes.

Basements

3.8 Because of the risk that a single stairway may be blocked by smoke from a fire in the basement or ground storey, the guidance in Section 2 paragraph 2.10, about basements in dwelling houses, applies equally to basement flats and maisonettes.

Balconies and flat roofs

3.9 The guidance in Section 2 paragraphs 2.5 and 2.6 on balconies and flat roofs of dwelling houses, applies equally to flats and maisonettes. In addition any balcony outside an alternative exit to a dwelling more than 4.5m above ground level should be a common balcony and meet the conditions in paragraph 3.15.

Provisions for escape from flats and maisonettes where the floor is not more than 4.5m above ground level

3.10 All rooms in the upper storey(s) should comply with Section 2 paragraph 2.7 if the design of the dwelling and the common means of escape does not follow the guidance in paragraphs 3.11 to 3.39. All rooms in the ground storey should comply with paragraph 2.8.

Additional provisions for flats and maisonettes with a floor more than 4.5m above ground level

Internal planning of flats

3.11 Three acceptable approaches (all of which should observe the restrictions concerning inner rooms given in paragraph 3.7) when planning a flat which has a floor at more than 4.5m above ground level are:

a. to provide a protected entrance hall which serves all habitable rooms, planned so that the travel distance from the entrance door to the door to any habitable room is 9m or less (see Diagram 7); or

b. to plan the flat so that the travel distance from the entrance door to any point in any of the habitable rooms does not exceed 9m and the cooking facilities are remote from the entrance door and do not prejudice the escape route from any point in the flat, (see Diagram 8); or

c. to provide an alternative exit from the flat, complying with paragraph 3.12.

Diagram 7   Flat where all habitable rooms have direct access to an entrance hall

See para 3.11(a)

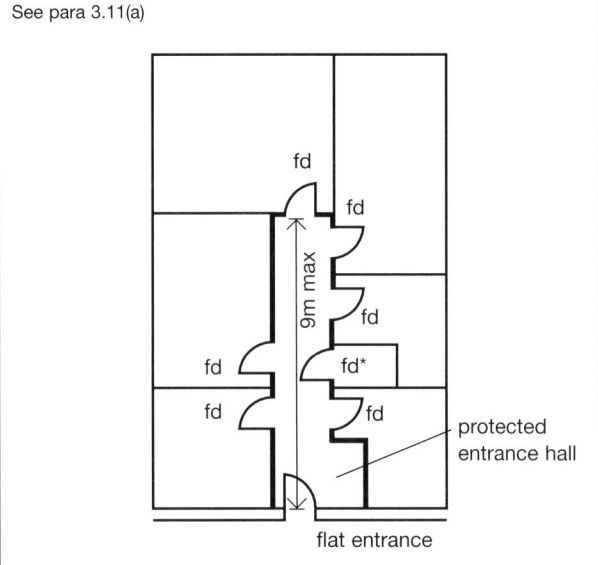

flat entrance

Note: Bathrooms need not have fire doors providing the bathroom is separated by fire-resisting construction from the adjacent rooms.

Key
fd   self-closing FD20 fire door
fd*   a cupboard door need not be self-closing
▬   30 minute fire-resisting construction around entrance hall

Diagram 8   Flat with restricted travel distance from furthest point to entrance

See para 3.11(b)

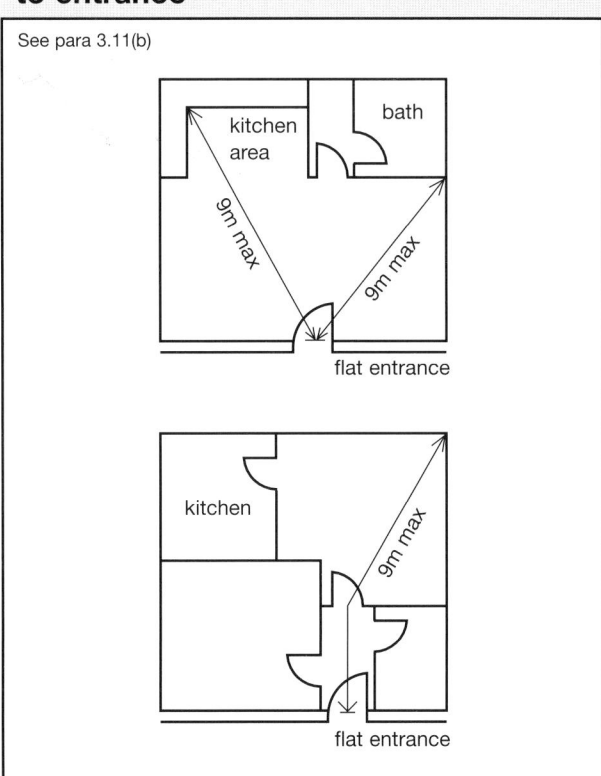

Diagram 9 **Flat with an alternative exit, but where all habitable rooms have no direct access to an entrance hall**

See para 3.12

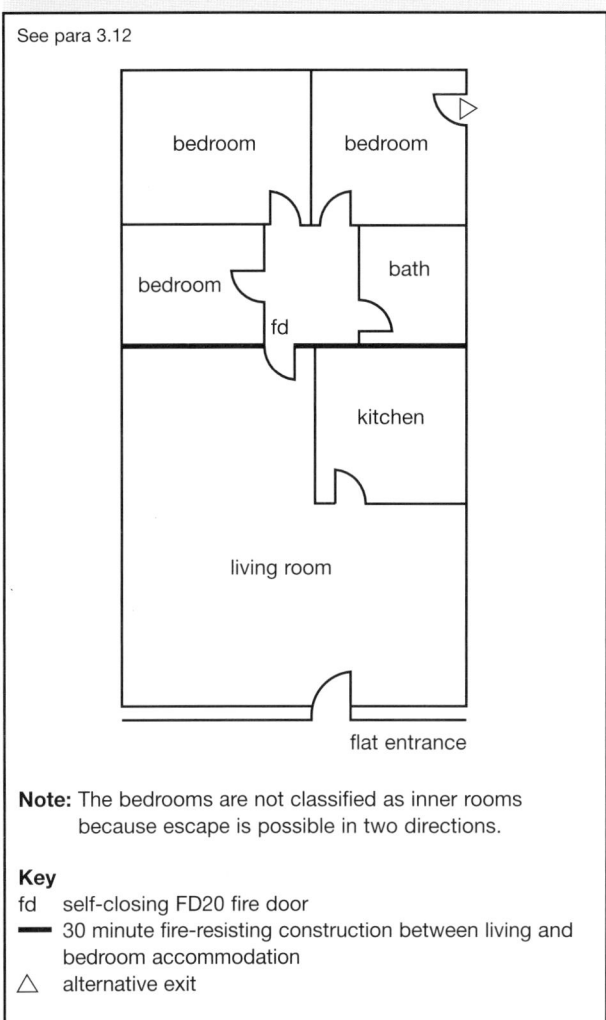

flat entrance

Note: The bedrooms are not classified as inner rooms because escape is possible in two directions.

Key
fd   self-closing FD20 fire door
▬   30 minute fire-resisting construction between living and bedroom accommodation
△   alternative exit

3.12 Where any flat has an alternative exit and the habitable rooms do not have direct access to the entrance hall, (see Diagram 9):

a.   the bedrooms should be separated from the living accommodation by fire-resisting construction and self-closing fire-door(s); and

b.   the alternative exit should be located in the part of the flat containing the bedroom(s).

Diagram 10 **Maisonette with alternative exits from each habitable room, except at entrance level**

See para 3.14(a)

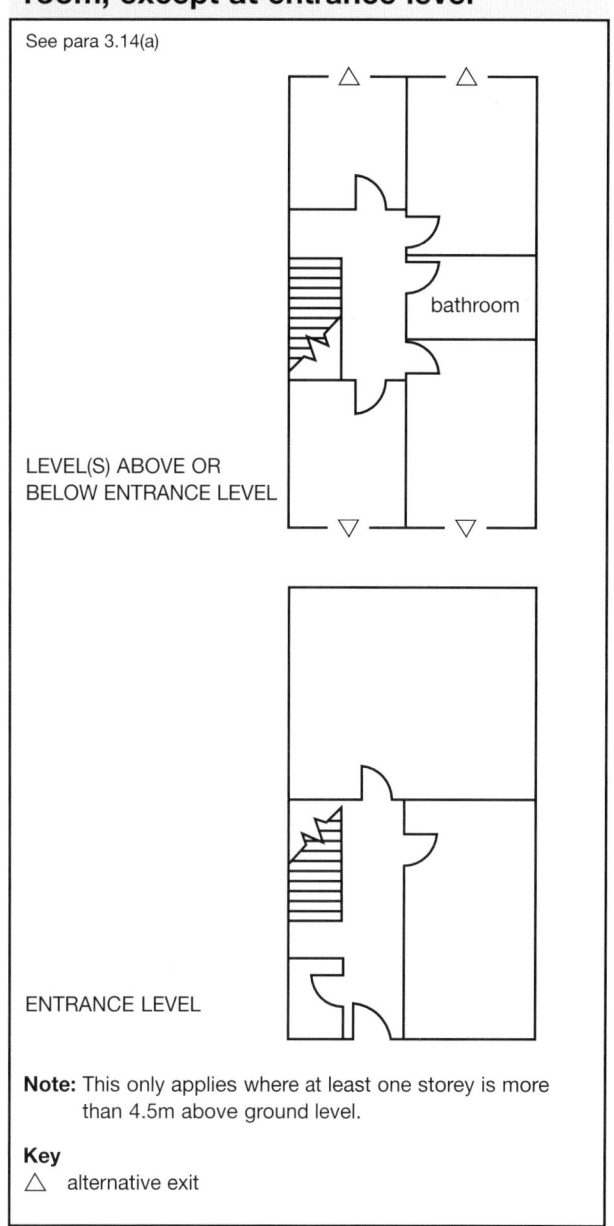

LEVEL(S) ABOVE OR BELOW ENTRANCE LEVEL

ENTRANCE LEVEL

Note: This only applies where at least one storey is more than 4.5m above ground level.

Key
△   alternative exit

Internal planning of maisonettes

3.13 A maisonette with an independent external entrance at ground level is similar to a dwellinghouse and means of escape should be planned on the basis of paragraphs 2.13 or 2.14 depending on the height of the top storey above ground level.

3.14 Two acceptable approaches to planning a maisonette, which does not have its own external entrance at ground level but has a floor at more than 4.5m above ground level, are:

a. to provide an alternative exit from each habitable room which is not on the entrance floor of the maisonette, see Diagram 10; or

b. to provide one alternative exit from each floor (other than the entrance floor), with a protected landing entered directly from all the habitable rooms on that floor, see Diagram 11.

Diagram 11 Maisonette with protected entrance hall and landing

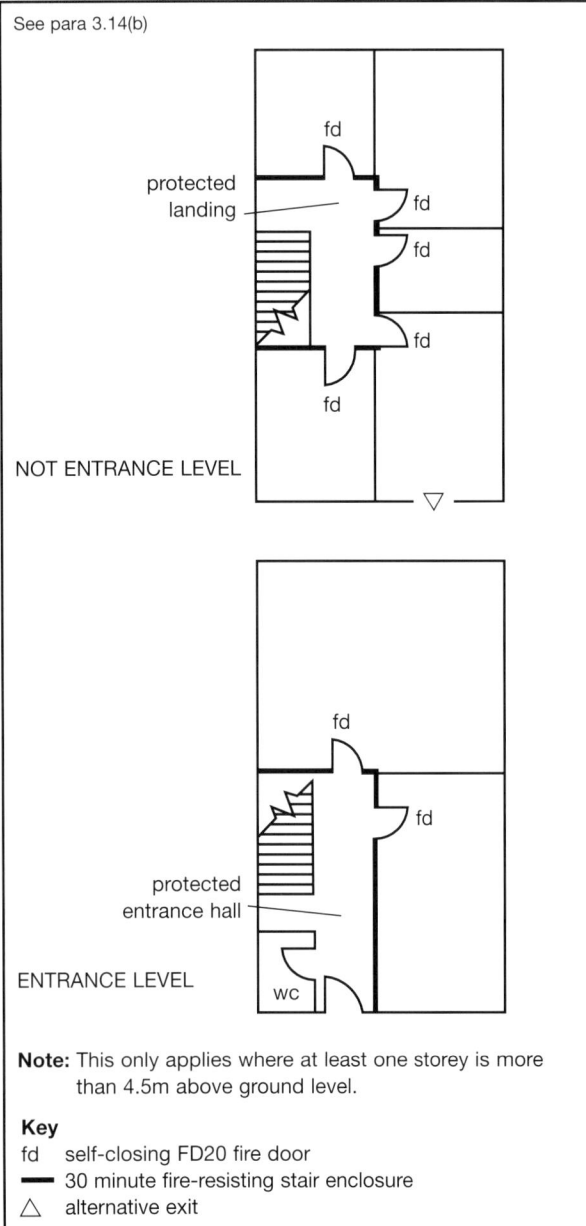

See para 3.14(b)

fd

protected landing

fd

fd

fd

fd

NOT ENTRANCE LEVEL

▽

fd

fd

protected entrance hall

ENTRANCE LEVEL

wc

Note: This only applies where at least one storey is more than 4.5m above ground level.

Key
fd   self-closing FD20 fire door
▬   30 minute fire-resisting stair enclosure
△   alternative exit

Alternative exits

3.15 To be effective, an alternative exit from a flat or maisonette should satisfy the following conditions:

a. be remote from the main entrance door to the dwelling; and

b. lead to a final exit or common stair by way of:

   i. a door onto an access corridor, access lobby or common balcony, or

   ii. an internal private stair leading to an access corridor, access lobby or common balcony at another level, or

   iii. a door onto a common stair, or

   iv. a door onto an external stair, or

   v. a door onto an escape route over a flat roof.

Note: Any such access to a final exit or common stair should meet the appropriate provisions dealing with means of escape in the common parts of the building (see paragraph 3.17).

Air circulation systems for heating, energy conservation or condensation control in flats and maisonettes with a floor more than 4.5m above ground level

3.16 With these types of systems, the following precautions are needed to avoid the possibility of the system allowing smoke or fire to spread into a protected entrance hall or landing.

a. Transfer grilles should not be fitted in any wall, door, floor or ceiling enclosing a protected entrance hall of a dwelling or protected stairway and landing of a maisonette.

b. All ductwork passing through the enclosure to a protected entrance hall or protected stairway and landing should be so fitted that all joints between the ductwork and the enclosure are fire-stopped.

c. Where ductwork is used to convey air into a protected entrance hall of the dwelling or protected stairway and landing within a maisonette through the enclosure of the protected hall or stairway, the return air from the protected hall or stairway should be ducted back to the plant.

d. Air and return air grilles or registers should be positioned at a height not exceeding 450mm above floor level.

e. A room thermostat for a ducted warm air heating system should be mounted at a height between 1370mm and 1830mm in an area from which air is drawn directly to the heating unit, and its maximum setting should not exceed 27°C.

f. Any system of mechanical ventilation which recirculates air should comply with paragraph 6.46.

Means of escape in the common parts of flats and maisonettes

3.17 The following paragraphs deal with means of escape from the entrance doors of dwellings to a final exit. They should be read in conjunction with the general provisions in Section 6.

Note: Paragraphs 3.18 to 3.48 are not applicable where the top floor is not more than 4.5m above ground level. However, attention is drawn to the provisions in paragraph 3.6 regarding sheltered housing, Section 6 regarding general provisions, Section 9 (B3) regarding the provision of compartment walls and protected shafts, and Section 17 (B5) regarding the provision of access for the fire service.

Number of escape routes

3.18 Every dwelling should have access to alternative escape routes so that a person confronted by the effects of an outbreak of fire in another dwelling can turn away from it and make a safe escape.

However, a single escape route from the dwelling entrance door is acceptable if either:

a.   the dwelling is situated in a storey served by a single common stair and:

   i.   every dwelling is separated from the common stair by a protected lobby or common corridor (see Diagram 12), and

   ii.   the travel distance limitations in Table 2, on escape in one direction only, are observed; or

b.   alternatively the dwelling is situated in a dead end part of a common corridor served by two (or more) common stairs, and the distance to the nearest common stair complies with the limitations in Table 2 on escape in one direction only (see Diagram 13).

Diagram 12 **Flats or maisonettes served by one common stair**

See paras 3.18(a) and 3.23

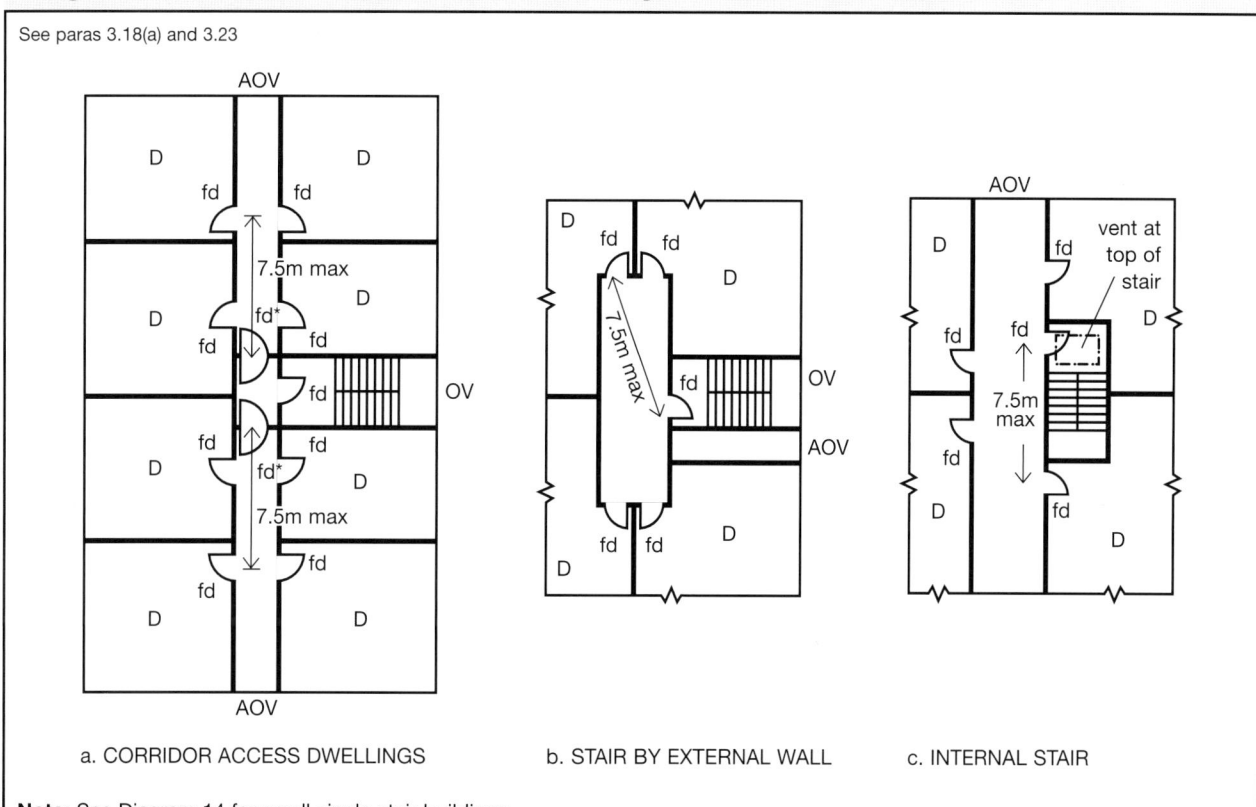

a. CORRIDOR ACCESS DWELLINGS   b. STAIR BY EXTERNAL WALL   c. INTERNAL STAIR

Note: See Diagram 14 for small single stair buildings.

Key

OV   openable vent at high level for fire service use (1.0m² minimum free area)
AOV   automatic opening vent at high level (1.5m² minimum free area)
D   dwelling
fd   self-closing FD30S fire door
fd*   self-closing FD20S fire door

Diagram 13 Flats or maisonettes served by more than one common stair

See paras 3.18(b) and 3.23

a. CORRIDOR ACCESS DWELLINGS ON ONE SIDE ONLY
maximum travel distance 30m

b. CORRIDOR ACCESS WITHOUT DEAD ENDS
maximum travel distance 30m

c. CORRIDOR ACCESS WITH DEAD ENDS
maximum travel distance 30m
(maximum from dead end 7.5m)
central door may be omitted
if maximum travel distance is no
more than 15m.

Key

OV	openable vent at high level for fire service use (1.0m² minimum free area)
AOV	automatic opening vent at high level (1.5m² minimum free area)
max TD	maximum travel distance
D	dwelling
fd	self-closing FD20S fire door (other doors shown are FD30S fire doors)

Diagram 14 Common escape route in small single stair building

See para 3.19

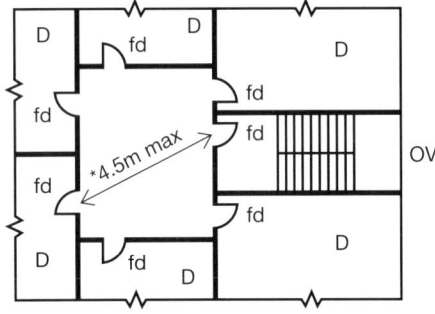

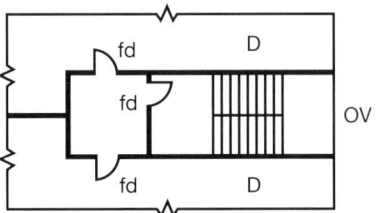

a. SMALL SINGLE STAIR BUILDING
*If an automatic opening vent is provided
in the lobby, the travel distance can be
increased to 7.5m maximum
(see Diagram 12, example b)

**b. SMALL SINGLE STAIR BUILDING
WITH NO MORE THAN 2 DWELLINGS
PER STOREY**
The door between stair and lobby should
be free from security fastenings.

If the dwellings have protected entrance
halls, the lobby between the common stair
and dwelling entrance is not essential

Notes:
1 The arrangements shown also apply to the top storey.
2 If the travel distance across the lobby in Diagram 14a exceeds 4.5m, Diagram 12b applies.
3 If there is one dwelling per storey in Diagram 14b, then the entrance door to the dwelling may form part of the stair enclosure, provided the dwelling has a protected entrance hall.

Key

▬	fire-resisting construction
OV	openable vent at high level for fire service use (1.0m² minimum free area). (It may be replaced by a vent over the stair.)
D	dwelling
fd	self-closing FD30S fire door

Small single-stair buildings

3.19 The provisions in paragraph 3.18 may be modified and a single stair, protected in accordance with Diagram 14, may be used provided that:

a. the top floor of the building is no more than 11m above ground level; and

b. there are no more than 3 storeys above the ground level storey;

c. the stair does not connect to a covered car park, except if the car park is open-sided (see paragraph 12.4 for meaning of open-sided car park); and

d. the stair does not serve ancillary accommodation unless:

   i. the storey containing the ancillary accommodation does not contain any dwellings, and

   ii. the ancillary accommodation is separated from the stair by a protected lobby, or protected corridor, which has not less than 0.4m² permanent ventilation or is protected from the ingress of smoke by a mechanical smoke control system.

Flats and maisonettes with balcony or deck access

3.20 The provisions of paragraph 3.18 may also be modified in the case of flats and maisonettes with balcony or deck approach. Guidance on these forms of development is set out in clause 13 of BS 5588 *Fire precautions in the design, construction and use of buildings*, Part 1: 1990 *Code of practice for residential buildings*.

Table 2 Limitations on distance of travel in common areas of flat and maisonette buildings
(see paragraph 3.21)

Maximum distance of travel (m) from dwelling entrance door to common stair, or to door to lobby in corridor-access single stair flats (Diagram 12a)

Escape in one direction only	Escape in more than one direction
7.5m (1)(2)	30m (2)(3)

Notes:

1. Reduced to 4.5m in the case shown in Diagram 14a.

2. Where all dwellings on a storey have independent alternative means of escape, the maximum distance of travel does not apply. However see paragraph 17.3 (B5) which specifies fire service access requirements.

3. For sheltered housing, see paragraph 3.6.

Common escape routes

Planning of common escape routes

3.21 Escape routes in the common areas should comply with the limitations on travel distance in Table 2. However there may be circumstances where some increase on these maximum figures will be reasonable.

Escape routes should be planned so that people do not have to pass through one stairway enclosure to reach another. However it is acceptable to pass through a protected lobby of one stairway in order to reach another.

Protection of common escape routes

3.22 To reduce the risk of a fire in a dwelling affecting the means of escape from other dwellings, and common parts of the building, the common corridors should be protected corridors.

The wall between each dwelling and the corridor should be a compartment wall (see Section 9).

Ventilation of common escape routes

3.23 Despite the provisions described in this Approved Document, it is probable that some smoke will get into a common corridor or lobby from a fire in a dwelling, if only because the entrance door will be open when the occupants escape.

There should therefore be some means of ventilating the common corridors/ lobbies to disperse smoke. (The ventilation also affords protection to the common stairs.)

a. In single stair buildings, other than small ones complying with Diagram 14, and in any dead-end portion of a building with more than one stair, the common corridor or lobby should be ventilated by an automatic opening ventilator, triggered by automatic smoke detection located in the space to be ventilated. The ventilator should have a free area of at least 1.5m², and be fitted with a manual override. (See also Diagram 12 and Diagram 13c);

b. In buildings with more than one stair, common corridors should extend at both ends to the external face of the building where there should be openable ventilators, which may operate automatically, for fire service use (see Diagram 13a and b). The free area of the ventilators should be at least 1.0m² at each end of the corridor.

Sub-division of common escape routes

3.24 A common corridor that connects two or more storey exits should be sub-divided by a self-closing fire door with, if necessary, any associated fire-resisting screen (see Diagram 13). The door(s) should be positioned so that smoke will not affect access to more than one stairway.

3.25 A dead end portion of a common corridor should be separated from the rest of the corridor by a self-closing fire door with, if necessary, any associated fire-resisting screen (see Diagram 12a and Diagram 13c).

Pressurization of common escape routes

3.26 Where the escape stairway and corridors/lobbies are protected by a smoke control system employing pressure differentials, the design should comply with BS 5588: Part 4 *Fire precautions in the design, construction and use of buildings, Code of practice for smoke control using pressure differentials*. (In such cases the cross corridor fire doors and the openable and automatically opening vents may be omitted).

Ancillary accommodation, etc

3.27 Stores and other ancillary accommodation should not be located within, or entered from, any protected lobby or protected corridor forming part of the only common escape route from a dwelling on the same storey as that ancillary accommodation.

Reference should be made to paragraphs 6.50 to 6.53 for special provisions for refuse chutes and storage areas.

Escape routes over flat roofs

3.28 If more than one escape route is available from a storey, or part of a building, one of those routes may be by way of a flat roof provided that it complies with the provisions in paragraph 6.35.

Note: Access to designs described in paragraph 3.45 may also be via a flat roof if the route over the roof complies with the provisions in paragraph 6.35.

Common stairs

Number of common stairs

3.29 As explained in paragraph 3.18 and paragraph 3.19 a single common stair can be acceptable in some cases, but otherwise there should be access to more than one common stair for escape purposes.

Width of common stairs

3.30 A stair of acceptable width for everyday use will be sufficient for escape purposes, but if it is also a firefighting stair, it should be at least 1100mm wide (see paragraph B1.xxviii for measurement of width).

Protection of common stairs

General

3.31 Common stairs need to have a satisfactory standard of fire protection if they are to fulfil their role as areas of relative safety during a fire evacuation. The provisions in paragraphs 3.32 to 3.44 below should be followed.

3.32 Stairs provide a potential route for fire spread from floor to floor. In Section 9 under the requirement of B3 to inhibit internal fire spread, there is guidance on the enclosure of stairs to avoid this. A stair may also serve as a firefighting stair in accordance with the requirement B5, in which case account will have to be taken of guidance in Section 18.

Enclosure of common stairs

3.33 Every common stair should be situated within a fire-resisting enclosure (ie it should be a protected stairway), to reduce the risk of smoke and heat making use of the stair hazardous.

3.34 The appropriate level of fire resistance is given in Appendix A Tables A1 and A2.

Exits from protected stairways

3.35 Every protected stairway should discharge:

a. directly to a final exit; or

b. by way of a protected exit passageway to a final exit.

Separation of adjoining protected stairways

3.36 Where two protected stairways (or exit passageways leading to different final exits) are adjacent, they should be separated by an imperforate enclosure.

Use of space within protected stairways

3.37 A protected stairway needs to be relatively free of potential sources of fire. Consequently, it should not be used for anything else, except a lift well or electricity meter(s). There are other provisions for lifts in paragraphs 6.39 to 6.45 and guidance on the installation of electricity meters is given in BS 5588 *Fire precautions in the design, construction and use of buildings*, Part 1 *Code of practice for residential buildings*.

Fire resistance and openings in external walls of protected stairways

3.38 The external enclosures to protected stairways should meet the provisions in paragraph 6.24.

Gas service and installation pipes in protected stairways

3.39 Gas service and installation pipes or associated meters should not be incorporated within a protected stairway unless the gas installation is in accordance with the requirements for installation and connection set out in the Pipelines Safety Regulations 1996, SI 1996 No 825 and the Gas Safety (Installation and Use) Regulations 1998 SI 1998 No 2451. (See also paragraph 9.41.)

Basement stairs

3.40 Because of their situation, basement stairways are more likely to be filled with smoke and heat than stairs in ground and upper storeys.

Special measures are therefore needed in order to prevent a basement fire endangering upper storeys. These are set out in the following two paragraphs.

3.41 If an escape stair forms part of the only escape route from an upper storey of a building (or part of a building) which is not a small building (see paragraph 3.19), it should not be continued down to serve any basement storey. The basement should be served by a separate stair.

3.42 If there is more than one escape stair from an upper storey of a building (or part of a building), only one of the stairs serving the upper storeys of the building (or part) need be terminated at ground level. Other stairs may connect with the basement storey(s) if there is a protected lobby, or a protected corridor between the stair(s) and accommodation at each basement level.

Stairs serving accommodation ancillary to flats and maisonettes

3.43 Except where described in paragraph 3.19, where a common stair forms part of the only escape route from a dwelling, it should not also serve any covered car park, boiler room, fuel storage space or other ancillary accommodation of similar fire risk on the same storey as that dwelling.

3.44 Any common stair which does not form part of the only escape route from a dwelling may also serve ancillary accommodation if it is separated from the ancillary accommodation by a protected lobby or a protected corridor.

If the stair serves an enclosed car park or place of special fire hazard, the lobby or corridor should have not less than 0.4m² permanent ventilation or be protected from the ingress of smoke by a mechanical smoke control system.

External escape stairs

3.45 If the building (or part of the building) is served by a single access stair, that stair may be external if it:

a. serves a floor not more than 6m above the ground level; and

b. meets the provisions in paragraph 6.25.

3.46 Where more than one escape route is available from a storey (or part of a building), some of the escape routes from that storey or part of the building may be by way of an external escape stair provided that there is at least one internal escape stair from every part of each storey (excluding plant areas) and the external stair(s):

a. serves a floor not more than 6m above either the ground level or a roof or podium which is itself served by an independent protected stairway; and

b. meets the provisions in paragraph 6.25.

Dwellings in mixed use buildings

Note: See also paragraph 5.4.

3.47 In buildings with not more than 3 storeys above the ground storey, stairs may serve both dwellings and other occupancies, provided that the stairs are separated from each occupancy by protected lobbies at all levels.

3.48 In buildings with more than 3 storeys above the ground storey, stairs may serve both dwellings and other occupancies provided that:

a. the dwelling is ancillary to the main use of the building and is provided with an independent alternative escape route;

b. the stair is separated from any other occupancies on the lower storeys by protected lobbies (at those storey levels).

   Note: The stair enclosure should have at least the same standard of fire resistance as stipulated in Table A2 for the elements of structure of the building (and take account of any additional provisions in Section 18 if it is a firefighting stair).

c. any automatic fire detection and alarm system with which the main part of the building is fitted also covers the dwelling;

d. any security measures should not prevent escape at all material times.

Note: Additional measures, including increased periods of fire resistance between the dwelling and the storage area may be required where fuels such as petrol and LPG are present. Guidance on this is referenced in BS 5588: Part 0 *Fire precautions in the design, construction and use of buildings, Guide to fire safety codes of practice for particular premises/applications.*

Section 4

DESIGN FOR HORIZONTAL ESCAPE – BUILDINGS OTHER THAN DWELLINGS

Introduction

4.1 The general principle to be followed when designing facilities for means of escape is that any person confronted by an outbreak of fire within a building can turn away from it and make a safe escape. This Section deals with the provision of means of escape from any point to the storey exit of the floor in question, for all types of building other than dwellinghouses, flats and maisonettes (for which refer to Sections 2 & 3). It should be read in conjunction with the guidance on the vertical part of the escape route in Section 5 and the general provisions in Section 6.

It should be noted that guidance in this Section is directed mainly at smaller, simpler types of buildings. Detailed guidance on the needs of larger, more complex or specialised buildings, can be found in the BS 5588 series of codes and elsewhere (see paragraph B1.xviii).

It should also be noted that although most of the information contained in this Section is related to general issues of design, special provisions apply to the layouts of certain institutional buildings (see paragraphs 4.29 onwards).

In the case of small shop, office, industrial, storage and other similar premises (ones with no storey larger than 280m^2 and having no more than 2 storeys plus a basement storey), the guidance in clause 10 of BS 5588: Part 11: 1997 *Fire precautions in the design, construction and use of buildings, Code of practice for shops, offices, industrial, storage and other similar buildings* may be followed instead of the provisions in this Section.

Escape route design

Number of escape routes and exits

4.2 The number of escape routes and exits to be provided depends on the number of occupants in the room, tier or storey in question, and the limits on travel distance to the nearest exit given in Table 3.

Note: It is only the distance to the nearest exit that should be so limited. Any other exits may be further away than the distances in Table 3.

4.3 In multi-storey buildings (see Section 5) more than one stair may be needed for escape, in which case every part of each storey will need to have access to more than one stair. This does not prevent areas from being in a dead-end condition provided that the alternative stair is accessible in case the first one is not usable.

4.4 In mixed use buildings, separate means of escape should be provided from any storeys (or parts of storeys) used for Residential or Assembly and Recreation purposes.

However, **see also paragraphs 3.47 onwards** which describe the circumstances under which a dwelling may be served by a stair which connects with other parts of a mixed use building.

Single escape routes and exits

4.5 In order to avoid occupants being trapped by fire or smoke, there should be alternative escape routes from all parts of the building.

However in the following situations a single route is acceptable:

a. parts of a floor from which a storey exit can be reached within the travel distance limit for travel in one direction set in Table 3 (but see also paragraph 4.6) provided that, in the case of places of assembly and bars, no one room in this situation has an occupant capacity of more than 60 people or 30 people if the building is in Institutional use (Purpose Group 2a). The calculation of capacity is described in B1.xxvi;

b. a storey (except one used for in-patient care in a hospital) with an occupant capacity of not more than 60 people, where the limits on travel in one direction only are satisfied (see Table 3).

Note: For schools, see paragraph 5.5(b).

Table 3 **Limitations on travel distance**

Purpose group	Use of the premises or part of the premises	Maximum travel distance (1) where travel is possible in:	
		one direction only (m)	more than one direction (m)
2(a)	Institutional (2)	9	18
2(b)	Other residential a. in bedrooms (3) b. in bedroom corridors c. elsewhere	9 9 18	18 35 35
3	Office	18	45
4	Shop and Commercial (4)	18 (5)	45
5	Assembly and Recreation a. buildings primarily for disabled people except schools b. schools c. areas with seating in rows d. elsewhere	9 18 15 18	18 45 32 45
6	Industrial (6)	25	45
7	Storage and other non-residential (6)	25	45
2-7	Place of special fire hazard (7)	9 (8)	18 (8)
2-7	Plant room or rooftop plant: a. distance within the room b. escape route not in open air (overall travel distance) c. escape route in open air (overall travel distance)	9 18 60	35 45 100

Notes:

1. The dimensions in the Table are travel distances. If the internal layout of partitions, fittings, etc is not known when plans are deposited, direct distances may be used for assessment. The direct distance is taken as 2/3rds of the travel distance.

2. If provision for means of escape is being made in a hospital or other health care building by following the detailed guidance in the relevant part of the Department of Health "Firecode", the recommendations about travel distances in the appropriate "Firecode" document should be followed.

3. Maximum part of travel distance within the room. (This limit applies within the bedroom (and any associated dressing room, bathroom or sitting room, etc) and is measured to the door to the protected corridor serving the room or suite. Sub-item (b) applies from that point along the bedroom corridor to a storey exit.)

4. Maximum travel distances within shopping malls are given in BS 5588: Part 10. Guidance on associated smoke control measures is given in a BRE report *Design methodologies for smoke and heat exhaust ventilation* (BR 368).

5. BS 5588: Part 10 applies more restrictive provisions to units with only one exit in covered shopping complexes.

6. In industrial and storage buildings the appropriate travel distance depends on the level of fire risk associated with the processes and materials being used. Control over the use of industrial buildings is exercised through the Fire Precautions Act. Attention is drawn to the guidance issued by the Home Office *Guide to fire precautions in existing places of work that require a fire certificate Factories Offices Shops and Railway Premises*. The dimensions given above assume that the premises will be of "normal" fire risk, as described in the Home Office guidance. If the building is high risk, as assessed against the criteria in the Home Office guidance, then lesser distances of 12m in one direction and 25m in more than one direction, would apply.

7. Places of special fire hazard are listed in the definitions in Appendix E.

8. Maximum part of travel distance within the room/area. Travel distance outside the room/area to comply with the limits for the purpose group of the building or part.

4.6 In many cases there will not be an alternative at the beginning of the route. For example, there may be only one exit from a room to a corridor, from which point escape is possible in two directions. This is acceptable provided that the overall distance to the nearest storey exit is within the limits for routes where there is an alternative, and the 'one direction only' section of the route does not exceed the limit for travel where there is no alternative, see Table 3. Diagram 15 shows an example of a dead-end condition in an open storey layout.

Diagram 15 Travel distance in dead-end condition

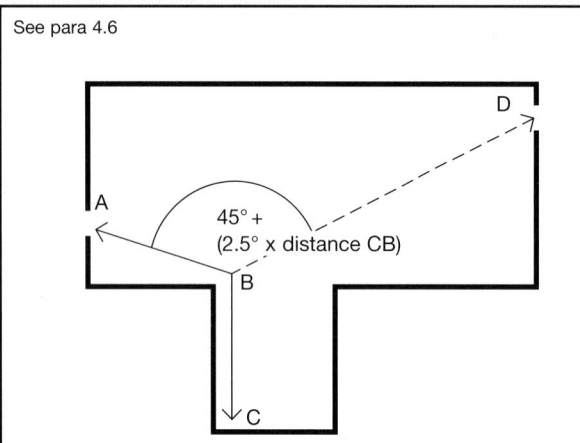

See para 4.6

Angle ABD should be at least 45° plus 2.5° for each metre travelled in one direction from point C. CBA or CBD (whichever is less) should be no more than the maximum distance of travel given for alternative routes, and CB should be no more than the maximum distance for travel where there are no alternative routes.

Number of occupants and exits

4.7 The figure used for the number of occupants will normally be that specified as the basis for the design. When the number of occupants likely to use a room, tier or storey is not known, the capacity should be calculated on the basis of the appropriate floor space factors. Guidance for this is set out in paragraph B1.xxvi and Table 1.

Table 4 gives the minimum number of escape routes and exits from a room, tier or storey according to the number of occupants. (This number is likely to be increased by the need to observe travel distances, and by other practical considerations.).

The width of escape routes and exits is the subject of paragraph 4.16.

Table 4 Minimum number of escape routes and exits from a room, tier or storey

Maximum number of persons	Minimum number of escape routes/exits
60	1
600	2
more than 600	3

Alternative escape routes

4.8 A choice of escape routes is of little value if they are all likely to be disabled simultaneously. Alternative escape routes should therefore satisfy the following criteria:

a. they are in directions 45° or more apart (see Diagram 16); or

b. they are in directions less than 45° apart, but are separated from each other by fire-resisting construction.

Diagram 16 Alternative escape routes

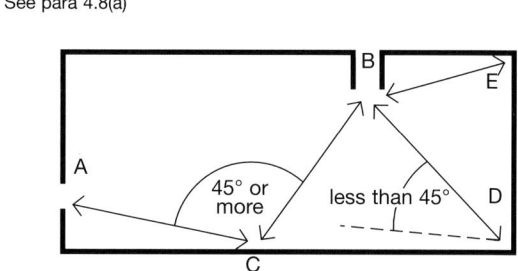

See para 4.8(a)

Alternative routes are available from C because angle ACB is 45° or more, and therefore CA or CB (whichever is the less) should be no more than the maximum distance for travel given for alternative routes.

Alternative routes are not available from D because angle ADB is less than 45° (therefore see Diagram 15). There is also no alternative route from E.

Inner rooms

4.9 A room from which the only escape route is through another room is called an inner room. It is at risk if a fire starts in the other room, called the access room.

Such an arrangement is only acceptable if the following conditions are satisfied:

a. the occupant capacity of the inner room should not exceed 60 (30 in the case of a building in purpose group 2a (Institutional));

b. the inner room should not be a bedroom;

c. the inner room should be entered directly off the access room;

d. the escape route from the inner room should not pass through more than one access room;

e. the travel distance from any point in the inner room to the exit(s) from the access room should not exceed the appropriate limit given in Table 3;

f. the access room should not be a place of special fire hazard and should be in the control of the same occupier; and

g. one of the following arrangements should be made:

   i. the enclosures (walls or partitions) of the inner room should be stopped at least 500mm below the ceiling, or

   ii. a suitably sited vision panel not less than 0.1m² should be located in the door or walls of the inner room, to enable occupants of the inner room to see if a fire has started in the outer room, or

   iii. the access room should be fitted with a suitable automatic fire detection and alarm system to warn the occupants of the inner room of the outbreak of a fire in the access room.

Planning of exits in a central core

4.10 Buildings with more than one exit in a central core should be planned so that storey exits are remote from one another, and so that no two exits are approached from the same lift hall, common lobby or undivided corridor, or linked by any of these. (See Diagram 18).

Access to storey exits

4.11 Any storey which has more than one escape stair, should be planned so that it is not necessary to pass through one stairway to reach another. However it would be acceptable to pass through one stairway's protected lobby to reach another stair.

Diagram 17 Inner room and access room

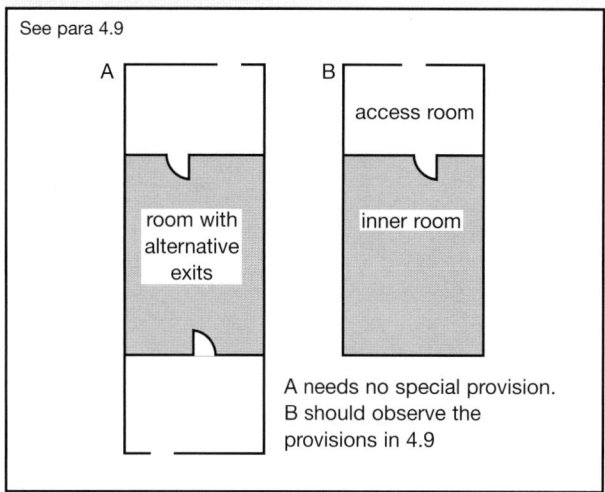

See para 4.9

A needs no special provision. B should observe the provisions in 4.9

Diagram 18 Exits in a central core

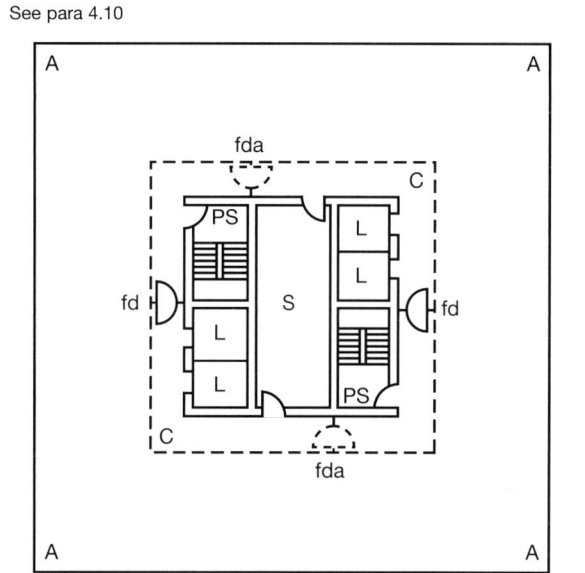

See para 4.10

Note: The doors at both ends of the area marked 'S' should be self-closing FD20S fire doors unless the area is sub-divided such that any fire in that area will not be able to prejudice both sections of corridor at the same time. If that area is a lift lobby, doors should be provided as shown in Figure 8 in BS 5588: Part 11: 1997.

Key
L   lift
S   services, toilets, etc.
fd self-closing FD20S fire doors
fda possible alternative position for fire door
C corridor off which accommodation opens
PS protected stairway
A accommodation (eg office space)

Separation of circulation routes from stairways

4.12 Unless the doors to a protected stairway and any associated exit passageway are fitted with an automatic release mechanism (see Appendix B, paragraph 3b), the stairway and any associated exit passageway should not form part of the primary circulation route between different parts of the building at the same level. This is because the self-closing fire doors are more likely to be rendered ineffective as a result of their constant use, or because some occupants may regard them as an impediment. For example the doors are likely to be wedged open or have their closers removed.

Storeys divided into different uses

4.13 Where a storey contains an area (which is ancillary to the main use of the building) for the consumption of food and/or drink by customers, then:

a. not less than two escape routes should be provided from each such area (except inner rooms which meet the provisions in 4.9); and

b. the escape routes from each such area should lead directly to a storey exit without entering any kitchen or similar area of high fire hazard.

Storeys divided into different occupancies

4.14 Where any storey is divided into separate occupancies (ie where there are separate ownerships or tenancies of different organisations):

a. the means of escape from each occupancy should not pass through any other occupancy; and

b. if the means of escape include a common corridor or circulation space, then either it should be a protected corridor or a suitable automatic fire detection and alarm system should be installed throughout the storey.

Height of escape routes

4.15 All escape routes should have a clear headroom of not less than 2m except in doorways.

Width of escape routes and exits

4.16 The width of escape routes and exits depends on the number of persons needing to use them. They should not be less than the dimensions given in Table 5. (Attention is also drawn to the guidance in Approved Document M *Access and facilities for disabled people*.)

4.17 Where the maximum number of people likely to use the escape route and exit is not known, the appropriate capacity should be calculated on the basis of the occupant capacity. Guidance is set out in paragraph B1.xxvi and Table 1.

4.18 Guidance on the spacing of fixed seating for auditoria is given in BS 5588: Part 6 *Fire precautions in the design, construction and use of buildings, Code of practice for assembly buildings*.

Table 5 **Widths of escape routes and exits** (1)

Maximum number of persons	Minimum width mm (2)(3)(4)
50	750 (5)
110	850
220	1050
more than 220	5 per person (6)

Notes:

1. In schools, the minimum width of corridors in pupil areas should be 1050mm (1600mm in dead-ends).
2. Refer to paragraph B1.xxviii on methods of measurement and width.
3. In order to follow the guidance in the Approved Document to Part M on minimum widths for areas accessible to disabled people, the widths given in the table may need to be increased.
4. Widths less than 1050mm should not be interpolated.
5. May be reduced to 530mm for gangways between fixed storage racking, other than in public areas of Purpose Group 4 (Shop & Commercial).
6. 5mm/person does not apply to an opening serving less than 220 persons.

Calculating exit capacity

4.19 If a storey has two or more storey exits it has to be assumed that a fire might prevent the occupants from using one of them. The remaining exit(s) need to be wide enough to allow all the occupants to leave quickly. Therefore when deciding on the total width of exits needed according to Table 5, the largest exit should be discounted. This may have implications for the width of stairs, because they should be at least as wide as any storey exit leading onto them. Although some stairs are not subject to discounting (see paragraphs 5.11 & 5.12), storey exits onto them will be.

4.20 The total number of persons which 2 or more available exits can accommodate is found by adding the maximum number of persons for each exit width. For example, 3 exits each 850mm wide will accommodate 3 x 110 = 330 persons (not the 510 persons accommodated by a single exit 2550mm wide).

Protected corridors

4.21 A corridor which serves a part of the means of escape in any of the following circumstances should be a protected corridor:

a. every corridor serving bedrooms;

b. every dead-end corridor (excluding recesses and extensions shown in Figures 10 and 11 in BS 5588: Part 11: 1997 *Code of practice for shops, offices, industrial, storage and other similar buildings*); and

c. any corridor common to two or more different occupancies (but see also paragraph 4.14).

Enclosure of corridors that are not protected corridors

4.22 Where a corridor that is used as a means of escape, but is not a protected corridor, is enclosed by partitions, those partitions provide some defence against the spread of smoke in the early stages of a fire, even though they may have no fire resistance rating. To maintain this defence the partitions should be carried up to the soffit of the structural floor above, or to a suspended ceiling, and openings into rooms from the corridor should be fitted with doors, which need not be fire doors. Open planning, while offering no impediment to smoke spread, has the compensation that occupants can become aware of a fire quickly.

Sub-division of corridors

4.23 If a corridor provides access to alternative escape routes, there is a risk that smoke will spread along it and make both routes impassable before all occupants have escaped.

To avoid this, every corridor more than 12m long which connects two or more storey exits, should be sub-divided by self-closing fire doors (and any necessary associated screens) so that the fire door(s) and any associated screen(s) are positioned approximately mid-way between the two storey exits to effectively safeguard the route from smoke (having regard to the layout of the corridor and to any adjacent fire risks).

4.24 If a dead-end portion of a corridor provides access to a point from which alternative escape routes are available, there is a risk that smoke from a fire could make both routes impassable before the occupants in the dead-end have escaped.

To avoid this, unless the escape stairway(s) and corridors are protected by a pressurization system complying with BS 5588: Part 4 *Fire precautions in the design, construction and use of buildings, Code of practice for smoke control using pressure differentials*, every dead-end corridor exceeding 4.5m in length should be separated by self-closing fire doors (together with any necessary associated screens) from any part of the corridor which:

a. provides two directions of escape (see Diagram 19(a)); or

b. continues past one storey exit to another (see Diagram 19(b)).

Cavity barriers

4.25 Additional measures to safeguard escape routes from smoke are given in Section 10 (B3), see items 5, 6 & 7 in Table 13.

Diagram 19 **Dead-end corridors**

See para 4.24

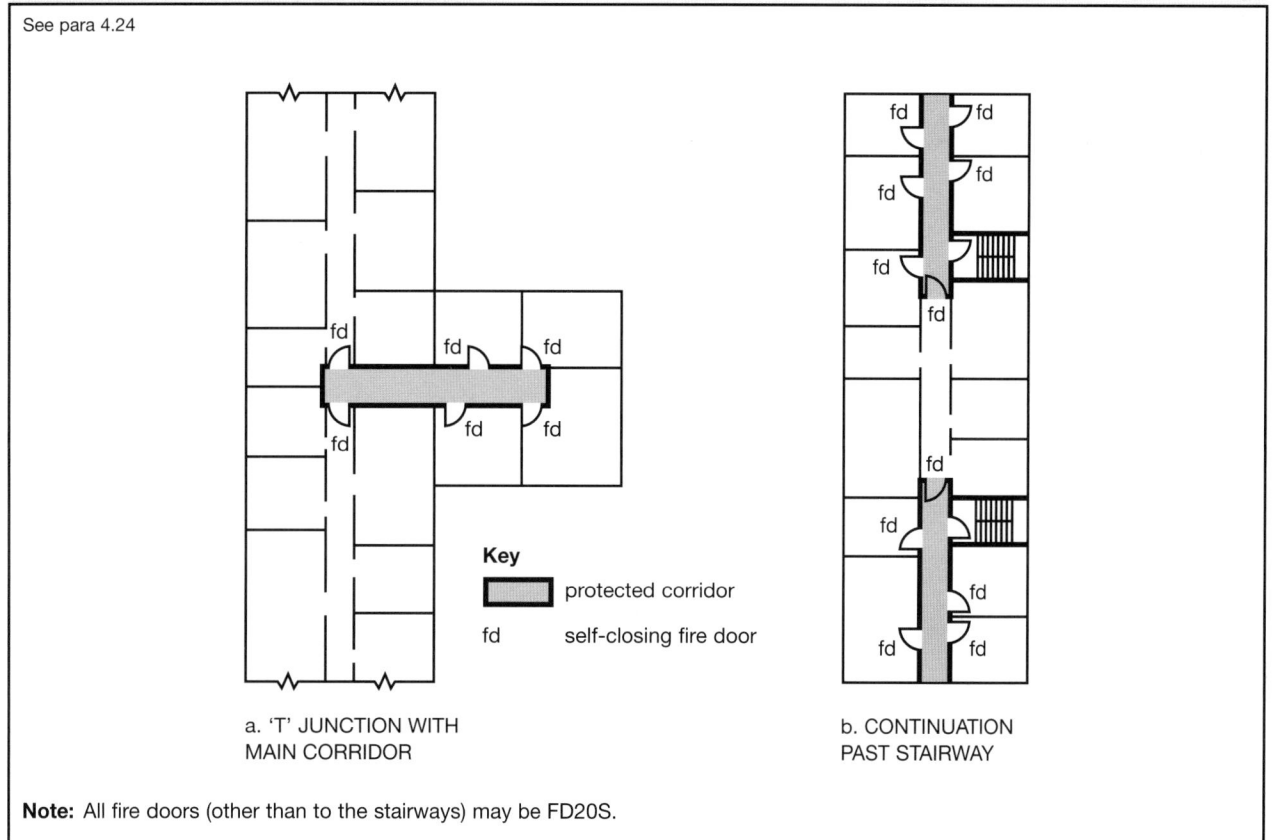

Key

protected corridor

fd   self-closing fire door

a. 'T' JUNCTION WITH MAIN CORRIDOR

b. CONTINUATION PAST STAIRWAY

Note: All fire doors (other than to the stairways) may be FD20S.

External escape routes

4.26 Guidance on the use of external escape stairs from buildings other than dwellings is given in paragraph 5.33.

4.27 Where an external escape route (other than a stair) is beside an external wall of the building, that part of the external wall within 1800mm of the escape route should be of fire-resisting construction, up to a height of 1100mm above the paving level of the route.

Escape over flat roofs

4.28 If more than one escape route is available from a storey, or part of a building, one of those routes may be by way of a flat roof, provided that:

a. the route does not serve an Institutional building, or part of a building intended for use by members of the public; and

b. it meets the provisions in paragraph 6.35.

Hospitals and other residential care premises of Purpose Group 2a

General

4.29 Paragraph B1.xx explains that the Department of Health "Firecode" documents should be used in the design of health care and other institutional premises, where the normal principles of evacuation are inappropriate.

Planning for progressive horizontal evacuation

4.30 The adoption of progressive horizontal evacuation may be of value in some other residential buildings. The following guidance is given for buildings to which the provisions of the "Firecode" documents are not applicable.

4.31 The concept of progressive horizontal evacuation allows progressive horizontal escape to be made by evacuating into adjoining compartments, or sub-divisions of compartments, in those areas used for inpatient care. The object is to provide a place of relative safety within a short distance, from which further evacuation can be made if necessary but under less pressure of time.

4.32 In planning a storey which is divided into compartments for progressive horizontal evacuation, the following conditions should be observed.

a. Adjoining compartments into which horizontal evacuation may take place should each have a floor area sufficient to accommodate not only their own occupants but also the occupants from the adjoining compartment. This should be calculated on the basis of the design occupancy of the compartments.

b. Each compartment should have at least one other escape route, independent of the route into the adjoining compartment, see Diagram 20. This other route may be by way of a third compartment, provided the exit from that compartment is independent of the exits from the other compartments.

Diagram 20 **Progressive horizontal evacuation**

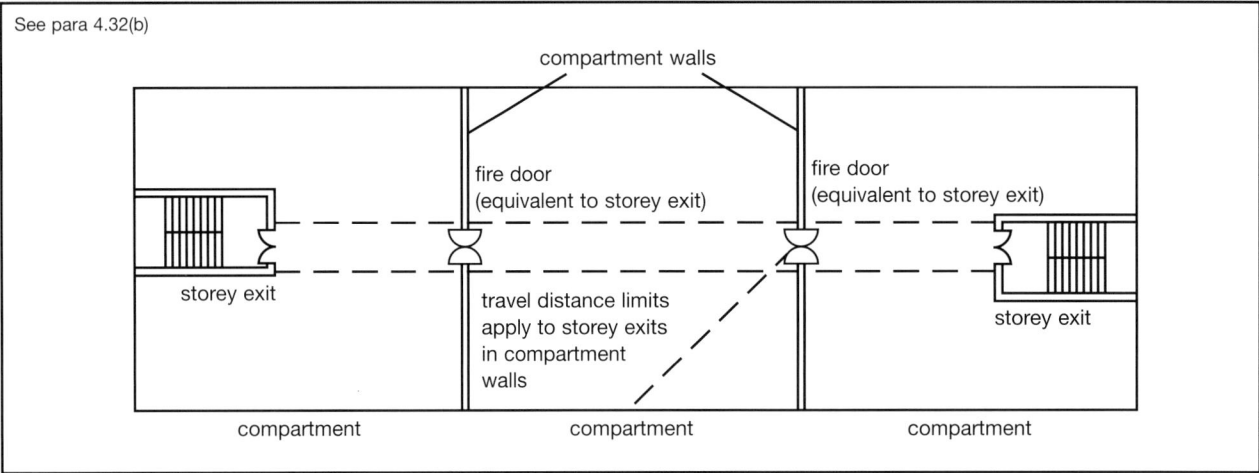

See para 4.32(b)

compartment walls

fire door (equivalent to storey exit)

fire door (equivalent to storey exit)

storey exit

storey exit

travel distance limits apply to storey exits in compartment walls

compartment

compartment

compartment

Section 5

DESIGN FOR VERTICAL ESCAPE – BUILDINGS OTHER THAN DWELLINGS

Introduction

5.1 An important aspect of means of escape in multi-storey buildings is the availability of a sufficient number of adequately sized and protected escape stairs. This Section deals with escape stairs and includes measures necessary to protect them in all types of building other than dwellinghouses, flats and maisonettes (for which see Sections 2 & 3).

This Section should be read in conjunction with the general provisions in Section 6.

Number of escape stairs

5.2 The number of escape stairs needed in a building (or part of a building) will be determined by:

a. the constraints imposed in Section 4 on the design of horizontal escape routes;

b. whether independent stairs are required in mixed occupancy buildings (see paragraph 5.4);

c. whether a single stair is acceptable (see paragraph 5.5); and

d. provision of adequate width for escape (see paragraph 5.6) while allowing for the possibility that a stair may have to be discounted because of fire or smoke (see paragraph 5.11).

5.3 In larger buildings, provisions for access for the fire service may apply in which case some escape stairs may also need to serve as firefighting stairs. The number of escape stairs may therefore be affected by provisions made in Section 18, paragraphs 18.7 and 18.8.

Mixed use buildings

5.4 Where a building contains storeys (or parts of storeys) in different purpose groups, it is important to consider the effect of one risk on another. A fire in a shop, or unattended office, could have serious consequences on, for example, a residential or hotel use in the same building. It is therefore important to consider whether completely separate routes of escape should be provided from each different use within the building or whether other effective means to protect common escape routes can be provided. (See paragraphs 3.47 and 3.48 for guidance where dwellings are served by a stair which connects with other parts of a mixed use building).

Single escape stairs

5.5 Provided that independent escape routes are not necessary from areas in different purpose groups in accordance with paragraph 5.4, the situations where a building (or part of a building) may be served by a single escape stair are:

a. from a basement which is allowed to have a single escape route in accordance with paragraph 4.5b;

b. from a building (other than small premises, see 5.5c) which has no storey with a floor level more than 11m above ground level, and in which every storey is allowed to have a single escape route in accordance with paragraph 4.5b (except that in schools, the storeys above the first floor level should only be occupied by adults);

Note: In a 2-storey school building (or part of a building) served by a single escape stair, there should be no more than 120 pupils plus supervisors on the first storey and no place of special fire hazard. Classrooms and stores should not open onto the stairway.

c. in the case of small premises, in situations where the recommendations of clause 10 of BS 5588 *Fire precautions in the design, construction and use of buildings*, Part 11: 1997 *Code of practice for shops, offices, industrial, storage and other similar buildings*, are followed.

Width of escape stairs

5.6 The width of escape stairs should:

a. be not less than the width(s) required for any exit(s) affording access to them;

b. conform with the minimum widths given in Table 6;

c. not exceed 1400mm if their vertical extent is more than 30m, unless it is provided with a central handrail (see notes 1 & 2 below); and

d. not reduce in width at any point on the way to a final exit.

Notes:

1. The 1400mm width has been given for stairs in tall buildings because research indicates that people prefer to stay within reach of a handrail, when making a prolonged descent, so that the centre part of a wider stair is little used and could be hazardous. Thus additional stair(s) may be needed.

2. Where a wider stair than 1400mm is provided with a central handrail, then the stair width on each side of the central handrail needs to be considered separately for the purpose of assessing stair capacity.

Table 6 Minimum widths of escape stairs

Situation of stair	Maximum number of of people served (1)	Minimum stair width (mm)
1a. In an institutional building (unless the stair will only be used by staff)	150	1000
1b. In an assembly building and serving an area used for assembly purposes (unless the area is less than 100m²)	220	1100
1c. In any other building and serving an area with an occupancy of more than 50	over 220	see Note (3)
2. Any stair not described above	50	800 (4)

Notes:

1. Assessed as likely to use the stair in a fire emergency.
2. BS 5588: Part 5 recommends that firefighting stairs should be at least 1100mm wide.
3. See Table 7 for sizing stairs for simultaneous evacuation, and Table 8 for phased evacuation.
4. In order to comply with the guidance in the Approved Document to Part M on minimum widths for areas accessible to disabled people, this may need to be increased to 1000mm.

5.7 If the resultant width of the stair is more than 1800mm, then for reasons of safety in use the guidance in Approved Document K *Protection from falling, collision and impact* is that, in public buildings the stair should have a central handrail. In such a case see Note 2 to paragraph 5.6.

5.8 Where an exit route from a stair also forms the escape route from the ground and/or basement storeys, the width may need to be increased accordingly.

Calculation of minimum stair width

General

5.9 Every escape stair should be wide enough to accommodate the number of persons needing to use it in an emergency. This width will depend on the number of stairs provided and whether the escape strategy is based on the simultaneous evacuation of the building, or part of the building (see paragraph 5.14) or phased evacuation (see paragraph 5.18).

5.10 As with the design of horizontal escape routes, where the maximum number of people needing to use the escape stairs is not known, the occupant capacity should be calculated on the basis of the appropriate floor space factors. Guidance for this is set out in paragraph B1.xxvi and Table 1.

Discounting of stairs

5.11 Whether phased or simultaneous evacuation is used, where two or more stairs are provided it should be assumed that one of them might not be available due to fire or smoke. It is therefore necessary to discount each stair in turn in order to ensure that the capacity of the remaining stair(s) is adequate for the number of persons needing to escape.

5.12 Two exceptions to the above discounting rules are if the escape stairs:

a. are protected by a smoke control system designed in accordance with BS 5588: Part 4 *Fire precautions in the design, construction and use of buildings, Code of practice for smoke control using pressure differentials*; or

b. are approached on each storey through a protected lobby (a protected lobby need not be provided on the topmost storey for the exception still to apply).

Note: Paragraph 5.24 identifies several cases where stairs need lobby protection.

In such cases the likelihood of a stair not being available is significantly reduced and it is not necessary to discount a stair. However, a storey exit needs to be discounted, see paragraph 4.19.

5.13 The stair discounting rule applies to a building fitted with a sprinkler system, unless the stairs are lobbied or protected by a smoke control system, as in 5.12.

Simultaneous evacuation

5.14 In a building designed for simultaneous evacuation, the escape stairs (in conjunction with the rest of the means of escape) should have the capacity to allow all floors to be evacuated simultaneously. In calculating the width of the stairs account is taken of the number of people temporarily housed in the stairways during the evacuation.

5.15 Escape based on simultaneous evacuation should be used for:

a. all stairs serving basements;

b. all stairs serving buildings with open spatial planning; and

c. all stairs serving Other Residential or Assembly and Recreation buildings.

Note: BS 5588 *Fire precautions in the design, construction and use of buildings*, Part 7: *Code of practice for the incorporation of atria in buildings*, includes designs based on simultaneous evacuation.

Table 7 Capacity of a stair for basements and for simultaneous evacuation of the building

No. of floors served				Maximum number of persons served by a stair of width:				
1000mm	1100mm	1200mm	1300mm	1400mm	1500mm	1600mm	1700mm	1800mm
1. 150	220	240	260	280	300	320	340	360
2. 190	260	285	310	335	360	385	410	435
3. 230	300	330	360	390	420	450	480	510
4. 270	340	375	410	445	480	515	550	585
5. 310	380	420	460	500	540	580	620	660
6. 350	420	465	510	555	600	645	690	735
7. 390	460	510	560	610	660	710	760	810
8. 430	500	555	610	665	720	775	830	885
9. 470	540	600	660	720	780	840	900	960
10. 510	580	645	710	775	840	905	970	1035

Notes:

1. The capacity of stairs serving more than 10 storeys may be obtained by using linear extrapolation.

2. The capacity of stairs not less than 1100mm wide may also be obtained by using the formula in paragraph 5.17.

3. Stairs with a rise of more than 30m should not be wider than 1400mm unless provided with a central handrail (see paragraph 5.6).

4. Stairs wider than 1800mm should be provided with a central handrail (see paragraph 5.7).

5.16 Where simultaneous evacuation is to be used, the capacity of stairs of widths from 1000 to 1800mm is given in Table 7.

5.17 As an alternative to using Table 7, the capacity of stairs 1100mm or wider (for simultaneous evacuation) can be derived from the formula:

P $= 200w + 50 (w - 0.3)(n - 1)$, or

w $= \dfrac{P + 15n - 15}{150 + 50n}$

where:

(P) is the number of people that can be served;

(w) is the width of the stair, in metres; and

(n) is the number of storeys served.

Notes:

1. Stairs with a rise of more than 30m should not be wider than 1400mm unless provided with a central handrail (see paragraph 5.6);

2. Separate calculations should be made for stairs/flights serving basement storeys and those serving upper storeys;

3. The population 'P' should be divided by the number of available stairs; and

4. The formula is particularly useful when determining the width of stairs serving a building (or part of a building) where the occupants are not distributed evenly – either within a storey or between storeys.

Worked examples:

A 14 storey building comprises 12 storeys of offices (ground + 11) with the top 2 storeys containing flats served by separate stairs. What is the minimum width needed for the stairs serving the office floors with a population of 1200 people (excluding the ground floor population which does not use the stairs), using simultaneous evacuation? Two stairs satisfy the travel distance limitations.

a. The population is distributed evenly.

As the top office storey is at a height greater than 18m, both stairs need added protection (see paragraph 5.24). Therefore if both stairs are entered at each level via a protected lobby, then both stairs can be assumed to be available (see paragraph 5.12).

P = 1200/2 = 600, n = 11
From the formula:
$600 = 200w + 50 (w - 0.3)(11 - 1)$
$600 = 200w + (50w - 15)(10)$
$600 = 200w + 500w - 150$
$750 = 700w$
w = 1070mm

Therefore both stairs should be at least 1070mm wide. But this needs to be increased to 1100mm as the formula applies to stairs 1100mm or wider (see paragraph 5.17).

b. The population is not distributed evenly (eg 1000 people occupy floors 1 to 9, and 200 occupy floors 10 & 11).

As the top office storey is at a height greater than 18m, both stairs need added protection (see paragraph 5.24). If both stairs are entered at each level via a protected lobby, then both stairs can be assumed to be available (see paragraph 5.12).

To find the width of

• the stairs serving floors 10 & 11:

P = 200/2 = 100, n = 2
From the formula:
100 = 200w + 50 (w – 0.3)(2 – 1)
100 = 200w + (50w – 15)(1)
100 = 200w + 50w – 15
115 = 250w
w = 460mm

Therefore both stairs between the 9th floor landing and the top floor should be at least 460mm. But this needs to be increased to 1100mm as the formula applies to stairs 1100mm or wider (see paragraph 5.17).

• the stairs serving floors 1 to 9:

P = 1200/2 = 600, n = 9
From the formula:
600 = 200w + 50 (w – 0.3)(9 – 1)
600 = 200w + (50w – 15)(8)
600 = 200w + 400w – 120
720 = 600w
w = 1200mm

Therefore both stairs between the 9th floor landing and the ground floor should be at least 1200mm wide.

Note: *In each example, the width will also be adequate when 1 storey exit is discounted in accordance with paragraph 4.19 and also with the need to comply with paragraph 5.6(a) (ie the stair is not less than the width found from Table 5).*

Phased evacuation

5.18 Where it is appropriate to do so, it may be advantageous to design stairs in high buildings on the basis of phased evacuation. In phased evacuation the first people to be evacuated are all those of reduced mobility and those on the storeys most immediately affected by the fire, usually the floor of fire origin and the floor above. Subsequently, if there is a need to evacuate more people, it is done two floors at a time. It is a method which cannot be used in every type of building, and it depends on the provision (and maintenance) of certain supporting facilities such as fire alarms. It does enable narrower stairs to be incorporated than would be the case if simultaneous evacuation were used, and has the practical advantage of reducing disruption in large buildings.

5.19 Phased evacuation may be used for any building provided it is not identified in paragraph 5.15 as needing simultaneous evacuation.

5.20 The following criteria should be satisfied in a building (or part of a building) that is designed on the basis of phased evacuation:

a. the stairways should be approached through a protected lobby or protected corridor at each storey, except a top storey;

b. the lifts should be approached through a protected lobby at each storey (see paragraph 6.42);

c. every floor should be a compartment floor;

d. if the building has a storey with a floor over 30m above ground level, the building should be protected throughout by an automatic sprinkler system meeting the relevant recommendations of BS 5306: Part 2 *Fire extinguishing installations and equipment on premises, Specification for sprinkler systems*, ie the relevant occupancy rating together with the additional requirements for life safety; this provision would not apply to any Purpose Group 1(a) (flats) part of a mixed use building;

e. the building should be fitted with an appropriate fire warning system, conforming to at least the L3 standard given in BS 5839: Part 1 *Fire detection and alarm systems for buildings, Code of practice for system design, installation and servicing*;

f. an internal speech communication system should be provided to permit conversation between a control point at fire service access level, and a fire warden on every storey. In addition, the recommendations relating to phased evacuation provided in BS 5839: Part 1 should be followed where it is deemed appropriate to install a voice alarm, this should be in accordance with BS 5839: Part 8 *Code of practice for the design, installation and servicing of voice alarm systems*.

5.21 The minimum width of stair needed when phased evacuation is used is given in Table 8. This table assumes a phased evacuation of not more than two floors at a time.

Table 8 Minimum width of stairs designed for phased evacuation

Maximum number of people in any storey	Stair width mm (1)
100	1000
120	1100
130	1200
140	1300
150	1400
160	1500
170	1600
180	1700
190	1800

Notes:

1. Stairs with a rise of more than 30m should not be wider than 1400mm unless provided with a central handrail (see paragraph 5.6).

2. As an alternative to using this table, provided that the minimum width of a stair is at least 1000mm, the width may be calculated from: [(P x 10) – 100]mm where P = the number

Worked example using Table 8

What is the minimum width needed for the stairs serving the office floors in the 14-storey building in the example to paragraph 5.17, assuming the population is distributed evenly.

As both stairs need to be entered at each level via a protected lobby (see paragraph 5.20a), then both stairs can be assumed to be available (see paragraph 5.12). Therefore:

- *number of persons per storey = 1200/11 = 109;*

- *each stair must be able to accommodate half the population of 1 storey (ie 109/2 = 55 persons)*

- *the width of 1 stair to accommodate 55 persons is 1000mm (maximum capacity 100 persons)*

- *thus both stairs need to be not less than 1000mm wide.*

This width will also be adequate when 1 storey exit is discounted in accordance with paragraph 4.19 and the need to comply with paragraph 5.6(a) (ie the stair is not less than the minimum width needed for 109 persons in Table 5).

Protection of escape stairs

General

5.22 Escape stairs need to have a satisfactory standard of fire protection if they are to fulfil their role as areas of relative safety during a fire evacuation. The guidance in paragraphs 5.23 to paragraph 5.31 should be followed to achieve this.

Enclosure of escape stairs

5.23 Every internal escape stair should be a protected stairway (ie it should be within a fire-resisting enclosure).

However an unprotected stair (eg an accommodation stair) may form part of an internal route to a storey exit or final exit, provided that the distance of travel and the number of people involved are very limited. For example, small premises described in clause 10 of BS 5588: Part 11: 1997 *Fire precautions in the design, construction and use of buildings, Code of practice for shops, offices, industrial, storage and other similar buildings* and raised storage areas.

There may be additional measures if the protected stairway is also a protected shaft (where it penetrates one or more compartment floors, see Section 9) or if it is a firefighting shaft (see Section 18).

Access lobbies and corridors

5.24 There are situations where an escape stair needs the added protection of a protected lobby or protected corridor. These are:

a. where the stair is the only one serving a building (or part of a building) which has more than one storey above or below the ground storey (except for small premises covered in paragraph 5.5c); or

b. where the stair serves any storey at a height greater than 18m; or

c. where the building is designed for phased evacuation (see paragraph 5.20a).

In these cases protected lobbies or protected corridors are needed at all levels, except the top storey, and at all basement levels; or

d. where the stair is a firefighting stair.

Lobbies are also needed where that option in paragraph 5.12 has been used to not discount one stairway when calculating stair widths.

An alternative that may be considered in (a) to (c) above is to use a smoke control system as described in paragraph 5.12.

5.25 A protected lobby should be provided between an escape stairway and a place of special fire hazard. In this case, the lobby should have not less than $0.4m^2$ permanent ventilation, or be protected from the ingress of smoke by a mechanical smoke control system.

Exits from protected stairways

5.26 Every protected stairway should discharge:

a. directly to a final exit; or

b. by way of a protected exit passageway to a final exit.

Note: Doors may be situated in the enclosures to the passageway, and lobbies will be needed to these doorways if the stairway is served by lobbies.

5.27 The exit from a protected stairway should meet the provisions in paragraph 6.31.

Separation of adjoining stairways

5.28 Where two protected stairways are adjacent, they, and any protected exit passageways linking them to final exits, should be separated by an imperforate enclosure.

Use of space within protected stairways

5.29 A protected stairway needs to be free of potential sources of fire. Consequently, facilities that may be incorporated in a protected stairway are limited to the following:

a. sanitary accommodation or washrooms, so long as the accommodation is not used as a cloakroom. A gas water heater or sanitary towel incinerator may be installed in the accommodation but not any other gas appliance;

b. a lift well may be included in a protected stairway, if it is not a firefighting stair;

c. a reception desk or enquiry office area at ground or access level, if it is not in the only stair serving the building or part of the building. The reception or enquiry office area should not be more than 10m² in area;

d. cupboards enclosed with fire-resisting construction, if it is not in the only stair serving the building or part of the building.

External walls of protected stairways

5.30 The external enclosures to protected stairways should meet the provisions in paragraph 6.24.

Gas service pipes in protected stairways

5.31 The reference to gas service pipes or associated meters set out in paragraph 3.39, also applies to buildings other than dwellings.

Basement stairs

5.32 The guidance on basement stairs in paragraphs 3.40 to 3.42, also applies to buildings other than dwellings

External escape stairs

5.33 If more than one escape route is available from a storey (or part of a building), some of the escape routes from that storey or part of the building may be by way of an external escape stair, provided that:

a. there is at least one internal escape stair from every part of each storey (excluding plant areas);

b. in the case of an Assembly and Recreation building, the route is not intended for use by members of the public; or

c. in the case of an Institutional building, the route serves only office or residential staff accommodation.

5.34 Where external stairs are acceptable as forming part of an escape route, they should meet the provisions in paragraph 6.25.

Section 6

GENERAL PROVISIONS COMMON TO BUILDINGS OTHER THAN DWELLINGHOUSES

Introduction

6.1 This section gives guidance on the construction and protection of escape routes generally, and on some services installations and other matters associated with the design of escape routes. It applies to all buildings other than dwellinghouses (refer to Section 2 for those).

It should therefore be read in conjunction with Section 3 (in respect of flats and maisonettes), and in conjunction with Sections 4 and 5 (in respect of other buildings).

Protection of escape routes

Fire resistance of enclosures

6.2 Details of fire resistance test criteria, and standards of performance, are set out in Appendix A. Generally a 30 minute standard is sufficient for the protection of means of escape. The exceptions to this are when greater fire resistance is required by the guidance on Requirements B3 or B5 or some other specific instance to meet Requirement B1, in Sections 4 and 5.

6.3 All walls, partitions and other enclosures that need to be fire-resisting to meet the provisions in this Approved Document (including roofs that form part of a means of escape), should have the appropriate performance given in Tables A1 and A2 of Appendix A.

6.4 Elements protecting a means of escape should meet any limitations on the use of glass (see paragraph 6.7).

Fire resistance of doors

6.5 Details of fire resistance test criteria, and standards of performance, are set out in Appendix B.

6.6 All doors that need to be fire-resisting to meet the provisions in this Approved Document should have the appropriate performance given in Table B1 of Appendix B.

Doors should also meet any limitations on the use of glass (see paragraph 6.7).

Fire resistance of glazed elements

6.7 Where glazed elements in fire-resisting enclosures and doors are only able to satisfy the relevant performance in terms of integrity, the use of glass is limited. These limitations depend on whether the enclosure forms part of a protected shaft (see Section 9) and the provisions set out in Appendix A, Table A4.

6.8 Where the relevant performance can be met in terms of both integrity and insulation, there is no restriction in this Approved Document on the use or amount of glass, but there are some restrictions on the use of glass in firefighting stairs and lobbies under the recommendations in clause 9 in BS 5588: Part 5: 1991 *Fire precautions in the design, construction and use of buildings, Code of practice for firefighting stairs and lifts* for robust construction (which is referred to in Section 18).

6.9 Attention is also drawn to the guidance on the safety of glazing in Approved Document N *Glazing – safety in relation to impact, opening and cleaning*.

Doors on escape routes

6.10 The time taken to negotiate a closed door can be critical in escaping. Doors on escape routes (both within and from the building) should therefore be readily openable, if undue delay is to be avoided. Accordingly the following provisions in paragraphs 6.11 to 6.18 should be met.

Door fastenings

6.11 In general, doors on escape routes (whether or not the doors are fire doors), should either not be fitted with lock, latch or bolt fastenings, or they should only be fitted with simple fastenings that can be readily operated from the side approached by people making an escape. The operation of these fastenings should be readily apparent and without the use of a key and without having to manipulate more than one mechanism. This is not intended to prevent doors being fitted with hardware to allow them to be locked when the rooms are empty. There may also be situations such as hotel bedrooms where locks may be fitted that are operated from the outside by a key and from the inside by a knob or lever etc.

6.12 In buildings where security on final exit doors is an important consideration, such as in some Assembly and Recreation or Shop and Commercial uses, panic bolts may be used. In non-residential buildings it may also be appropriate to accept on some final exit doors locks for security that are used only when the building is empty. In these cases the emphasis for the safe use of these locks must be placed on management procedures.

6.13 Guidance about door closing and 'hold open' devices for fire doors is given in Appendix B.

Direction of opening

6.14 The door of any doorway or exit should, if reasonably practicable, be hung to open in the direction of escape, and should always do so if the number of persons that might be expected to use the door at the time of a fire is more than 60.

Note: With respect to industrial activities where there is a very high fire risk with potential for rapid fire growth, there will be a requirement for the door to open in the direction of escape for lower numbers than 60.

Amount of opening and effect on associated escape routes

6.15 All doors on escape routes should be hung to open not less than 90 degrees, and with a swing that is clear of any change of floor level, other than a threshold or single step on the line of the doorway (see paragraph 6.21) and does not reduce the effective width of any escape route across a landing.

6.16 A door that opens towards a corridor or a stairway should be sufficiently recessed to prevent its swing from encroaching on the effective width of the stairway or corridor.

Vision panels in doors

6.17 Vision panels are needed where doors on escape routes sub-divide corridors, or where any doors are hung to swing both ways, but note also the provision in Approved Document M *Access and facilities for disabled people*, concerning vision panels in doors across accessible corridors and passageways and the provisions for the safety of glazing in Approved Document N *Glazing – safety in relation to impact, opening and cleaning*.

Revolving and automatic doors

6.18 Revolving doors, automatic doors and turnstiles can obstruct the passage of persons escaping. Accordingly, they should not be placed across escape routes unless:

a. they are to the required width and are automatic doors and either they:

i. are arranged to fail safely to outward opening from any position of opening, or

ii. are provided with a monitored failsafe system for opening the doors if the mains supply fails, or

iii. they fail safely to the open position in the event of power failure; or

b. non-automatic swing doors of the required width should be provided immediately adjacent to the revolving or automatic door or turnstile.

Stairs

Construction of escape stairs

6.19 The flights and landings of every escape stair should be constructed of materials of limited combustibility in the following situations:

a. if it is the only stair serving the building, or part of the building, unless the building is of two or three storeys and is in Purpose Group 1(a) or Purpose Group 3;

b. if it is within a basement storey (this does not apply to a private stair in a maisonette);

c. if it serves any storey having a floor level more than 18m above ground or access level;

d. if it is external, except in the case of a stair that connects the ground floor or paving level with a floor or flat roof not more than 6m above or below ground level. (There is further guidance on external escape stairs in paragraph 6.25); or

e. if it is a firefighting stair (see Section 18).

Note: In satisfying the above conditions combustible materials may be added to the upper surface of these stairs (except in the case of firefighting stairs).

6.20 There is further guidance on the construction of firefighting stairs in Section 18. Dimensional constraints on the design of stairs generally, to meet requirements for safety in use, are given in Approved Document K, *Protection from falling, collision and impact*.

Single steps

6.21 Single steps may cause falls and should only be used on escape routes where they are prominently marked. A single step on the line of a doorway is acceptable.

Helical stairs, spiral stairs and fixed ladders

6.22 Helical stairs, spiral stairs (but not for pupils in schools) and fixed ladders may form part of an escape route subject to the following restrictions.

a. helical and spiral stairs should be designed in accordance with BS 5395: Part 2 *Stairs, ladders and walkways, Code of practice for the design of helical and spiral stairs* and, if they are intended to serve members of the public, should be a type E (public) stair, in accordance with that standard;

b. fixed ladders should not be used as a means of escape for members of the public, and should only be intended for use in circumstances where it is not practical to provide a conventional stair, for example as access to plant rooms that are not normally occupied. Fixed ladders should be constructed of non-combustible materials.

6.23 Guidance on the design of helical and spiral stairs, and fixed ladders, from the aspect of safety in use, is given in Approved Document K *Protection from falling, collision and impact*.

External walls of protected stairways

6.24 With some configurations of external wall, a fire in one part of a building could subject the external wall of a protected stairway to heat (for example, where the two are adjacent at an internal angle in the facade as shown in Diagram 21). If the external wall of the protected stairway has little fire resistance, there is a risk that this could prevent the safe use of the stair. Therefore, if:

a. a protected stairway projects beyond, or is recessed from, or is in an internal angle of, the adjoining external wall of the building; then

b. the distance between any unprotected area in the external enclosures to the building and any unprotected area in the enclosure to the stairway should be at least 1800mm (see Diagram 21).

External escape stairs

6.25 Where an external escape stair is provided in accordance with paragraph 3.45, paragraph 3.46 or paragraph 5.33, it should meet the following provisions:

a. all doors giving access to the stair should be fire-resisting and self-closing, except that a fire-resisting door is not required at the head of any stair leading downwards where there is only one exit from the building onto the top landing;

b. any part of the external envelope of the building within 1800mm of (and 9m vertically below), the flights and landings of an external escape stair should be of fire-resisting construction, except that the 1800mm dimension may be reduced to 1100mm above the top level of the stair if it is not a stair up from a basement to ground level (see Diagram 22);

c. there is protection by fire-resisting construction for any part of the building (including any doors) within 1800mm of the escape route from the stair to a place of safety, unless there is a choice of routes from the foot of the stair that would enable the people escaping to avoid exposure to the effects of the fire in the adjoining building; and

d. any stair more than 6m in vertical extent is protected from the effects of adverse weather conditions. (This should not be taken to imply a full enclosure. Much will depend on the location of the stair and the degree of protection given to the stair by the building itself); and

e. glazing in areas of fire-resisting construction mentioned above should also be fire-resisting (integrity but not insulation) and fixed shut.

Diagram 21 **External protection to protected stairways**

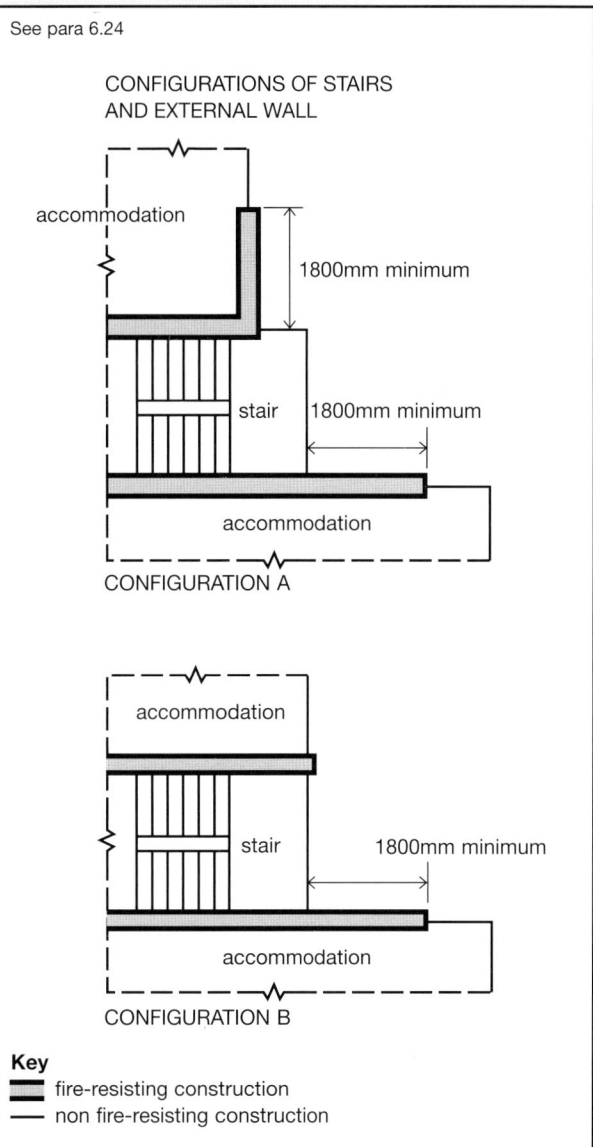

See para 6.24

CONFIGURATIONS OF STAIRS AND EXTERNAL WALL

accommodation

1800mm minimum

stair | 1800mm minimum

accommodation

CONFIGURATION A

accommodation

stair | 1800mm minimum

accommodation

CONFIGURATION B

Key
▬ fire-resisting construction
— non fire-resisting construction

Diagram 22 Fire resistance of areas adjacent to external stairs

See para 6.25

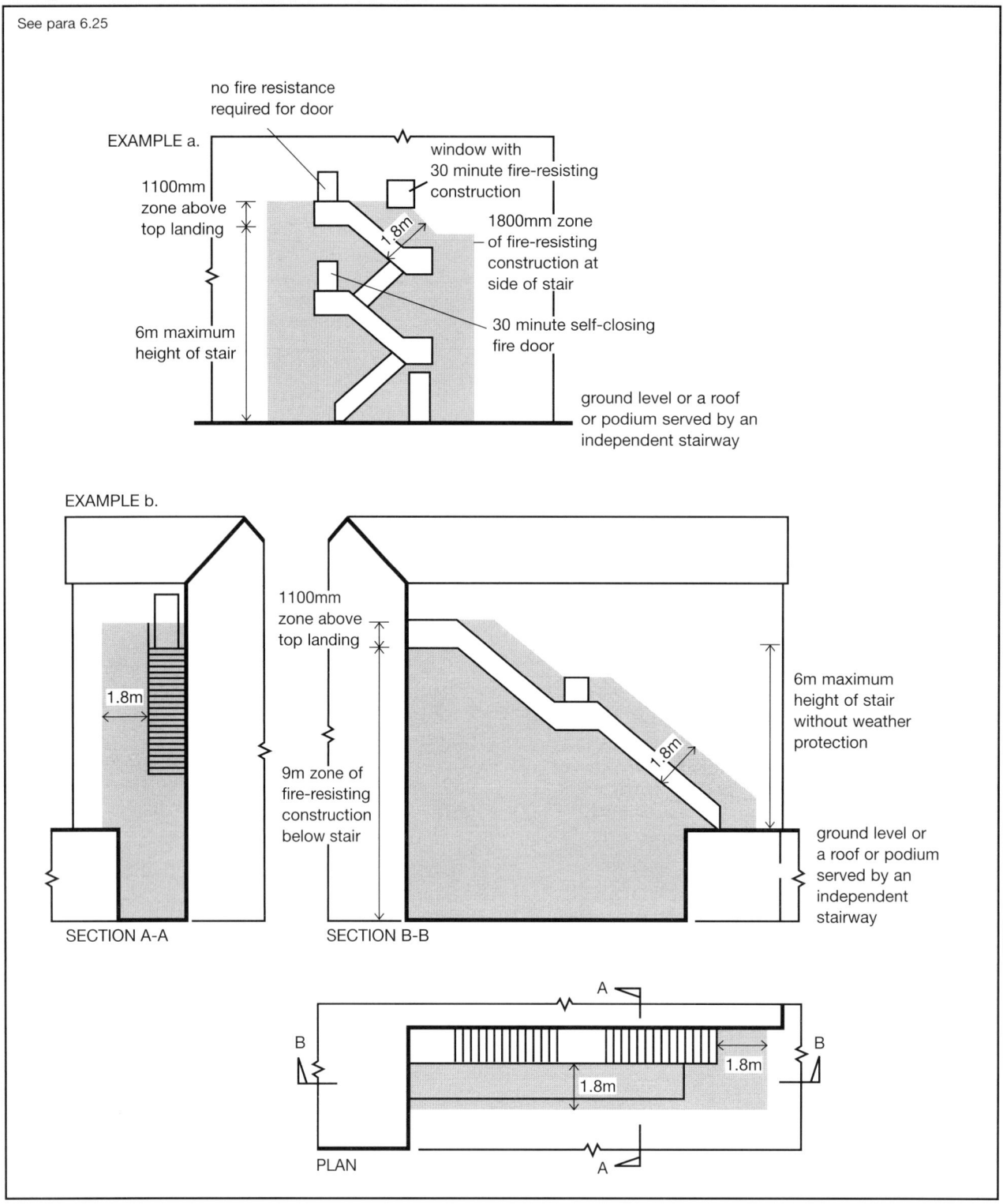

EXAMPLE a.

no fire resistance required for door

window with 30 minute fire-resisting construction

1100mm zone above top landing

1800mm zone of fire-resisting construction at side of stair

1.8m

30 minute self-closing fire door

6m maximum height of stair

ground level or a roof or podium served by an independent stairway

EXAMPLE b.

1.8m

1100mm zone above top landing

9m zone of fire-resisting construction below stair

6m maximum height of stair without weather protection

1.8m

ground level or a roof or podium served by an independent stairway

SECTION A-A

SECTION B-B

1.8m

1.8m

PLAN

General

Headroom in escape routes

6.26 All escape routes should have a clear headroom of not less than 2m and there should be no projection below this height (except for door frames).

Floors of escape routes

6.27 The floorings of all escape routes (including the treads of steps, and surfaces of ramps and landings) should be chosen to minimise their slipperiness when wet.

Ramps and sloping floors

6.28 Where a ramp forms part of an escape route it should meet the provisions in Approved Document M *Access and facilities for disabled people*.

6.29 Any sloping floor or tier should be constructed with a pitch of not more than 35° to the horizontal.

6.30 Further guidance on the design of ramps and associated landings, and on aisles and gangways in places where there is fixed seating, from the aspect of safety in use, is given in Approved Document K *Protection from falling, collision and impact*, and in Approved Document M *Access and facilities for disabled people*. The design of means of escape in places with fixed seating is dealt with in Section 4 by reference to BS 5588: Part 6 *Fire precautions in the design, construction and use of buildings, Code of practice for places of assembly*.

Final exits

6.31 Final exits need to be dimensioned and sited to facilitate the evacuation of persons out of and away from the building. Accordingly, they should be not less in width than the minimum width required for the escape route(s) they serve and should also meet the conditions in the following paragraphs 6.32 to 6.34.

6.32 Final exits should be sited to ensure rapid dispersal of persons from the vicinity of the building so that they are no longer in danger from fire and smoke. Direct access to a street, passageway, walkway or open space should be available. The route clear of the building should be well defined, and if necessary suitably guarded.

6.33 Final exits need to be apparent to persons who may need to use them. This is particularly important where the exit opens off a stair that continues down, or up, beyond the level of the final exit.

6.34 Final exits should be sited so that they are clear of any risk from fire or smoke in a basement (such as the outlets to basement smoke vents, see Section 19), or from openings to transformer chambers, refuse chambers, boiler rooms and similar risks.

Escape routes over flat roofs

6.35 Where an escape route over a flat roof is provided in accordance with paragraph 3.28 or paragraph 4.28, it should meet the following provisions:

a. the roof should be part of the same building from which escape is being made;

b. the route across the roof should leads to a storey exit or external escape route;

c. the part of the roof forming the escape route and its supporting structure, together with any opening within 3m of the escape route, should be fire-resisting (see Appendix A Table A1); and

d. the route should be adequately defined and guarded by walls and/or protective barriers which meet the provisions in Approved Document K, *Protection from falling, collision and impact.*

Lighting of escape routes

6.36 All escape routes should have adequate artificial lighting. Routes and areas listed in Table 9 should also have escape lighting which illuminates the route if the main supply fails.

Lighting to escape stairs should be on a separate circuit from that supplying any other part of the escape route.

Standards for the installation of a system of escape lighting are given in BS 5266: Part 1 *Emergency lighting. Code of practice for the emergency lighting of premises other than cinemas and certain other specified premises used for entertainment*, and CP 1007 *Maintained lighting for cinemas*.

Table 9 Provisions for escape lighting

Purpose group of the building or part of the building	Areas requiring escape lighting
1. Residential	All common escape routes (1), except in 2-storey flats
2. Office, Shop and Commercial (2) Industrial, Storage, Other non-residential	a. Underground or windowless accommodation b. Stairways in a central core or serving storey(s) more than 18m above ground level c. Internal corridors more than 30m long d. Open-plan areas of more than 60m²
3. Shop and Commercial (3) and car parks to which the public are admitted	All escape routes (1) (except in shops of 3 or fewer storeys with no sales floor more than 280m² provided that the shop is not a restaurant or bar)
4. Assembly and Recreation	All escape routes (1), and accommodation except for: a. accommodation open on one side to view sport or entertainment during normal daylight hours b. parts of school buildings with natural light and used only during normal school hours
5. Any purpose group	a. Windowless toilet accommodation with a floor area not more than 8m² b. All toilet accommodation with a floor area over 8m² c. Electricity and generator rooms d. Switch room/battery room for emergency lighting system e. Emergency control room

Notes:

1. Including external escape routes.
2. Those parts of the premises to which the public are not admitted.
3. Those parts of the premises to which the public are admitted.

Exit signs

6.37 Except in dwellings, every escape route (other than those in ordinary use) should be distinctively and conspicuously marked by emergency exit sign(s) of adequate size complying with the Health and Safety (Safety signs and signals) Regulations 1996. In general, signs containing symbols or pictograms which conform to BS 5499: Part 1 *Fire safety signs, notices and graphic symbols, Specification for fire safety signs*, satisfy these regulations. In some buildings additional signs may be needed to meet requirements under other legislation.

Note: Advice on fire safety signs, including emergency escape signs, is given in an HSE publication: *Safety Signs and Signals: Guidance on Regulations*.

Protected power circuits

6.38 Where it is critical for electrical circuits to be able to continue to function during a fire, protected circuits are needed. A protected circuit for operation of equipment in the event of fire should consist of cable meeting the requirements for classification as CWZ in accordance with BS 6387 *Specification for performance requirements for cables required to maintain circuit integrity under fire conditions*. It should follow a route selected to pass only through parts of the building in which the fire risk is negligible and should be separate from any circuit provided for another purpose.

Lifts

Evacuation lifts

6.39 In general it is not appropriate to use lifts when there is a fire in the building because there is always the danger of people being trapped in a lift that has become immobilised as a result of the fire. However, in some circumstances a lift may be provided as part of a management plan for evacuating disabled persons. In such cases the lift installation needs to be appropriately sited and protected, and needs to contain a number of safety features that are intended to ensure that the lift remains usable for evacuation purposes during the fire.

Guidance on the necessary measures is given in BS 5588: Part 8 *Fire precautions in the design, construction and use of buildings, Code of practice for means of escape for disabled people*.

Fire protection of lift installations

6.40 Because lifts connect floors, there is the possibility that they may prejudice escape routes. To safeguard against this, the following conditions in paragraphs 6.41 to 6.45 should be met.

6.41 Lifts, such as wall-climber or feature lifts which rise within a large volume such as a mall or atrium, and do not have a conventional well, may be at risk if they run through a smoke reservoir. In which case care is needed to maintain the integrity of the smoke reservoir, and protect the occupants of the lift.

6.42 Lift wells should be either:

a. contained within the enclosures of a protected stairway, or

b. be enclosed throughout their height with fire-resisting construction if they are sited so as to prejudice the means of escape.

A lift well connecting different compartments should form a protected shaft (see Section 9).

In buildings designed for phased or progressive horizontal evacuation, where the lift well is not contained within the enclosures of a protected stairway, the lift entrance should be separated from the floor area on every storey by a protected lobby.

6.43 In basements and enclosed car parks the lift should be approached only by a protected lobby (or protected corridor) unless it is within the enclosure of a protected stairway.

This is also the case in any storey that contains high fire risk areas, if the lift also delivers directly into corridors serving sleeping accommodation. Examples of fire risk areas in this context are kitchens, lounges and stores.

6.44 A lift shaft should not be continued down to serve any basement storey if it is:

a. in a building (or part of a building) served by only one escape stair, and smoke from a basement fire would be able to prejudice the escape routes in the upper storeys; or

b. within the enclosures to an escape stair which is terminated at ground level.

6.45 Lift machine rooms should be sited over the lift well whenever possible. If the lift well is within a protected stairway which is the only stairway serving the building (or part of the building), then if the machine room cannot be sited above the lift well it should be located outside the stairway (to avoid smoke spread from a fire in the machine room).

Mechanical ventilation and air conditioning systems

6.46 Any system of mechanical ventilation should be designed to ensure that in a fire the air movement in the building is directed away from protected escape routes and exits, or that the system (or an appropriate section of it) is closed down. In the case of a system which recirculates air, it should meet the relevant recommendation for recirculating distribution systems in BS 5588: Part 9 *Fire precautions in the design, construction and use of buildings, Code of practice for ventilation and air conditioning ductwork*, in terms of its operation under fire conditions.

6.47 Guidance on the use of mechanical ventilation in a place of assembly is given in BS 5588: Part 6 *Code of practice for places of assembly*.

6.48 Where a pressure differential system is installed, ventilation and air conditioning systems in the building should be compatible with it when operating under fire conditions.

6.49 Guidance on the design and installation of mechanical ventilation and air conditioning plant is given in BS 5720 *Code of practice for mechanical ventilation and air conditioning in buildings*, and on ventilation and air conditioning ductwork in BS 5588: Part 9 *Code of practice for ventilation and air conditioning ductwork*.

Note: Paragraphs 9.41 and 11.10 also deal with ventilation and air-conditioning ducts.

Refuse chutes and storage

6.50 Refuse storage chambers, refuse chutes and refuse hoppers should be sited and constructed in accordance with BS 5906 *Code of practice for storage and on-site treatment of solid waste from buildings*.

6.51 Refuse chutes and rooms provided for the storage of refuse should:

a. be separated from other parts of the building by fire-resisting construction; and

b. not be located within protected stairways or protected lobbies.

6.52 Rooms containing refuse chutes, or provided for the storage of refuse, should be approached either directly from the open air or by way of a protected lobby provided with not less than $0.2m^2$ of permanent ventilation.

6.53 Access to refuse storage chambers should not be sited adjacent to escape routes or final exits, or near to windows of dwellings.

Shop store rooms

6.54 Fully enclosed walk-in store rooms in shops (unless provided with an automatic fire detection and alarm system or fitted with sprinklers) should be separated from retail areas with fire-resisting construction (see Appendix A, Table A1) if they are sited so as to prejudice the means of escape.

The Requirement

This Approved Document deals with the following
Requirement from Part B of Schedule 1 to the
Building Regulations 2000.

Requirement	*Limits on application*

Internal fire spread (linings)
 	
B2.-(1) To inhibit the spread of fire within the building, the internal linings shall-
 	
  (a) adequately resist the spread of flame over their surfaces; and
 	
  (b) have, if ignited, either a rate of heat release or a rate of fire growth, which is reasonable in the circumstances.
 	
(2) In this paragraph "internal linings" means the materials or products used in lining any partition, wall, ceiling or other internal structure.	

Guidance

Performance

In the Secretary of State's view the requirement of B2 will be met if the spread of flame over the internal linings of the building is restricted by making provision for them to have low rates of surface spread of flame, and in some cases to have a low rate of heat release, so as to limit the contribution that the fabric of the building makes to fire growth. In relation to the European fire tests and classification system, the requirement of B2 will be met if the heat released from the internal linings is restricted by making provision for them to have a resistance to ignition and a rate of fire growth which are reasonable in the circumstances.

The extent to which this is necessary is dependent on the location of the lining.

Introduction

Fire spread and lining materials

B2.i The choice of materials for walls and ceilings can significantly affect the spread of a fire and its rate of growth, even though they are not likely to be the materials first ignited.

It is particularly important in circulation spaces where linings may offer the main means by which fire spreads, and where rapid spread is most likely to prevent occupants from escaping.

Several properties of lining materials influence fire spread. These include the ease of ignition and the rate at which the lining material gives off heat when burning. The guidance relating to the European fire tests and classification provides for control of internal fire spread through control of these properties. This document does not give detailed guidance on other properties such as the generation of smoke and fumes.

Floors and stairs

B2.ii The provisions do not apply to the upper surfaces of floors and stairs because they are not significantly involved in a fire until well developed, and thus do not play an important part in fire spread in the early stages of a fire that are most relevant to the safety of occupants.

However, it should be noted that the construction of some stairs and landings is controlled under Section 6, paragraph 6.19, and in the case of firefighting stairs, Section 18, paragraph 18.11.

Other controls on internal surface properties

B2.iii There is also guidance on the control of flame spread inside buildings in two other Sections. In Section 10 there is guidance on surfaces exposed in concealed spaces above fire-protecting suspended ceilings, and in Section 11 on enclosures to above ground drainage system pipes.

Note: External flame spread is dealt with in Sections 13-15; the fire behaviour of insulating core panels used for internal structures is dealt with in Appendix F.

Furniture and fittings

B2.iv Furniture and fittings can have a major effect on fire spread but it is not possible to control them through Building Regulations, and they are not dealt with in this Approved Document. Fire characteristics of furniture and fittings may be controlled in some buildings under legislation that applies to a building in use, such as licensing conditions.

Classification of performance

B2.v Appendix A describes the different classes of performance and the appropriate methods of test (see paragraphs 7-20).

The National classifications used are based on tests in BS 476: Fire tests on building materials and structures, namely Part 6: *Method of test for fire propagation for products* and Part 7: *Method of test to determine the classification of the surface spread of flame of products*. However, Part 4: *Non-combustibility test for materials* and Part 11: *Method for assessing the heat emission from building products* are also used as one method of meeting Class 0. Other tests are available for classification of thermoplastic materials if they do not have the appropriate rating under BS 476 Part 7 and three ratings, referred to as TP(a) rigid and TP(a) flexible and TP(b), are used.

The European classifications are described in BS EN 13501-1:2002, *Fire classification of construction products and building elements,* Part 1-*Classification using data from reaction to fire tests*. They are based on a combination of four European test methods, namely:

BS EN ISO 1182:2002, *Reaction to fire tests for building products – Non combustibility test;*

BS EN ISO 1716:2002, *Reaction to fire tests for building products – Determination of the gross calorific value;*

BS EN 13823:2002, *Reaction to fire tests for building products – Building products excluding floorings exposed to the thermal attack by a single burning item*; and

BS EN ISO 11925-2:2002, *Reaction to fire tests for building products, Part 2-Ignitability when subjected to direct impingement of flame.*

For some building products, there is currently no generally accepted guidance on the appropriate procedure for testing and classification in accordance with the harmonised European fire tests. Until such a time that the appropriate European test and classification methods for these building products are published classification may only be possible using existing national test methods.

Table A8, in Appendix A, gives typical performance ratings which may be achieved by some generic materials and products.

Section 7

WALL AND CEILING LININGS

Classification of linings

7.1 Subject to the variations and specific provisions described in paragraphs 7.2 to 7.17 below, the surface linings of walls and ceilings should meet the following classifications:

Table 10 Classification of linings

Location	National class (1)	European class (1)(3)(4)
Small rooms (2) of area not more than: a) 4m² in residential accommodation; b) 30m² in non-residential accommodation	3	D-s3, d2
Domestic garages of area not more than 40m²		
Other rooms (2) (including garages)	1	C-s3, d2
Circulation spaces within dwellings		
Other circulation spaces, including the common areas of flats and maisonettes	0	B-s3, d2

Notes:

1   see paragraph B2.v.

2   for meaning of room, see definition in Appendix E.

3   the National classifications do not automatically equate with the equivalent classifications in the European column, therefore products cannot typically assume a European class, unless they have been tested accordingly.

4   when a classification includes "s3, d2", this means that there is no limit set for smoke production and/or flaming droplets/particles.

Definition of walls

7.2 For the purpose of the performance of wall linings, a wall includes:

a. the surface of glazing (except glazing in doors); and

b. any part of a ceiling which slopes at an angle of more than 70° to the horizontal.

But a wall does not include:

c. doors and door frames;

d. window frames and frames in which glazing is fitted;

e. architraves, cover moulds, picture rails, skirtings and similar narrow members; and

f. fireplace surrounds, mantle shelves and fitted furniture.

Definition of ceilings

7.3 For the purposes of the performance of ceiling linings, a ceiling includes:

a. the surface of glazing;

b. any part of a wall which slopes at an angle of 70° or less to the horizontal.

But a ceiling does not include:

c. trap doors and their frames;

d. the frames of windows or rooflights (see Appendix E) and frames in which glazing is fitted;

e. architraves, cover moulds, picture rails, exposed beams and similar narrow members.

Variations and special provisions

Walls

7.4 Parts of walls in rooms may be of a poorer performance than specified in paragraph 7.1 (but not poorer than Class 3 (National class) or Class D-s3, d2 (European class)) provided the total area of those parts in any one room does not exceed one half of the floor area of the room, subject to a maximum of 20m² in residential accommodation, and 60m² in non-residential accommodation.

Fire-protecting suspended ceilings

7.5 A suspended ceiling can contribute to the overall fire resistance of a floor/ceiling assembly. Such a ceiling should satisfy paragraph 7.1. It should also meet the provisions of Appendix A, Table A3.

Fire-resisting ceilings

7.6 Cavity barriers are needed in some concealed floor or roof spaces (see Section 10), however this need can be reduced by the use of a fire-resisting ceiling below the cavity. Such a ceiling should comply with Diagram 35.

Rooflights

7.7 Rooflights should meet the relevant classification in 7.1. However plastic rooflights with at least a Class 3 rating may be used where 7.1 calls for a higher standard, provided the limitations in Table 11 below and in Table 18 are observed.

Note: No guidance is currently possible on the performance requirements in the European fire tests as there is no generally accepted test and classification procedure.

Special applications

7.8 Air supported structures should comply with the recommendations given in BS 6661 *Guide for the design, construction and maintenance of single-skin air supported structures.*

7.9 Any flexible membrane covering a structure (other than an air supported structure) should comply with the recommendations given in Appendix A of BS 7157 *Method of test for ignitability of fabrics used in the construction of large tented structures*.

7.10 Guidance on the use of PTFE-based materials for tension-membrane roofs and structures is given in a BRE Report *Fire safety of PTFE-based materials used in buildings* (BR 274, BRE 1994).

Thermoplastic materials

General

7.11 Thermoplastic materials (see Appendix A, paragraph 17) which cannot meet the performance given in Table 10, can nevertheless be used in windows, rooflights and lighting diffusers in suspended ceilings if they comply with the provisions described in paragraphs 7.12 to 7.16 below. Flexible thermoplastic material may be used in panels to form a suspended ceiling if it complies with the guidance in paragraph 7.17. The classifications used in paragraphs 7.12 to 7.17, Table 11 and Diagram 24 are explained in Appendix A, paragraph 20.

Note: No guidance is currently possible on the performance requirements in the European fire tests as there is no generally accepted test and classification procedure.

Windows and internal glazing

7.12 External windows to rooms (though not to circulation spaces) may be glazed with thermoplastic materials, if the material can be classified as a TP(a) rigid product.

Internal glazing should meet the provisions in paragraph 7.1 above.

Notes:

1. A "wall" does not include glazing in a door (see paragraph 7.2);

2. Attention is drawn to the guidance on the safety of glazing in Approved Document N *Glazing – safety in relation to impact, opening and cleaning*.

Rooflights

7.13 Rooflights to rooms and circulation spaces (with the exception of protected stairways) may be constructed of a thermoplastic material if:

a. the lower surface has a TP(a) (rigid) or TP(b) classification;

b. the size and disposition of the rooflights accords with the limits in Table 11 and with the guidance to B4 in Table 19.

Lighting diffusers

7.14 The following provisions apply to lighting diffusers which form part of a ceiling, and are not concerned with diffusers of light fittings which are attached to the soffit of, or suspended beneath a ceiling (see Diagram 23).

Lighting diffusers are translucent or open-structured elements that allow light to pass through. They may be part of a luminaire or used below rooflights or other sources of light.

7.15 Thermoplastic lighting diffusers should not be used in fire-protecting or fire-resisting ceilings, unless they have been satisfactorily tested as part of the ceiling system that is to be used to provide the appropriate fire protection.

7.16 Subject to the above paragraphs, ceilings to rooms and circulation spaces (but not protected stairways) may incorporate thermoplastic lighting diffusers if the following provisions are observed.

a. Wall and ceiling surfaces exposed within the space above the suspended ceiling (other than the upper surfaces of the thermoplastic panels) should comply with the general provisions of paragraph 7.1, according to the type of space below the suspended ceiling.

b. If the diffusers are of classification TP(a) (rigid), there are no restrictions on their extent.

c. If the diffusers are of classification TP(b), they should be limited in extent as indicated in Table 11 and Diagram 24.

Suspended or stretched-skin ceilings

7.17 The ceiling of a room may be constructed either as a suspended or stretched skin membrane from panels of a thermoplastic material of the TP(a) flexible classification, provided that it is not part of a fire-resisting ceiling. Each panel should not exceed 5m^2 in area and should be supported on all its sides.

Diagram 23 **Lighting diffuser in relation to ceiling**

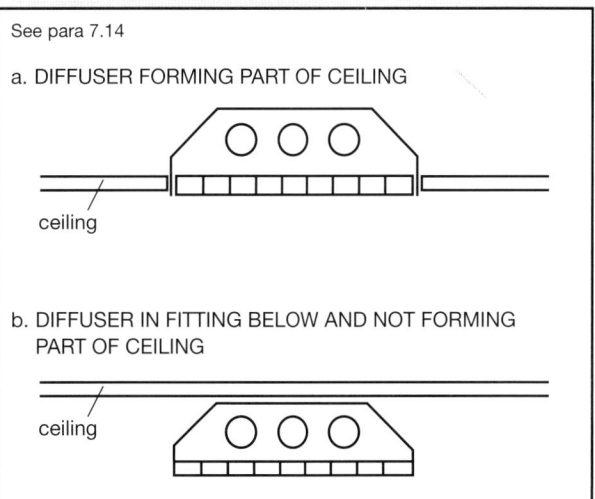

See para 7.14

a. DIFFUSER FORMING PART OF CEILING

ceiling

b. DIFFUSER IN FITTING BELOW AND NOT FORMING PART OF CEILING

ceiling

Table 11 Limitations applied to thermoplastic rooflights and lighting diffusers in suspended ceilings and Class 3 plastic rooflights

Minimum classification of lower surface	Use of space below the diffusers or rooflight	Maximum area of each diffuser panel or rooflight (1) (m²)	Max total area of diffuser panels and rooflights as percentage of floor area of the space in which the ceiling is located (%)	Minimum separation distance between diffuser panels or rooflights (1) (m)
TP(a)	any except protected stairway	No limit (2)	No limit	No limit
Class 3 (3) or TP(b)	rooms	5	50 (4)(5)	3 (5)
	circulation spaces except protected stairways	5	15 (4)	3

Notes:

1. Smaller panels can be grouped together provided that the overall size of the group and the space between one group and any others satisfies the dimensions shown in Diagram 24.

2. Lighting diffusers of TP(a) flexible rating should be restricted to panels of not more than 5 sq.m each, see paragraph 7.17

3. There are no limits on Class 3 material in small rooms.

4. The minimum 3m separation specified in Diagram 24 between each 5m² must be maintained. Therefore, in some cases it may not also be possible to use the maximum percentage quoted.

5. Class 3 rooflights to rooms in industrial and other non-residential purpose groups may be spaced 1800mm apart provided the rooflights are evenly distributed and do not exceed 20% of the area of the room.

Diagram 24 Layout restrictions on Class 3 plastics rooflights, TP(b) rooflights and TP(b) lighting diffusers

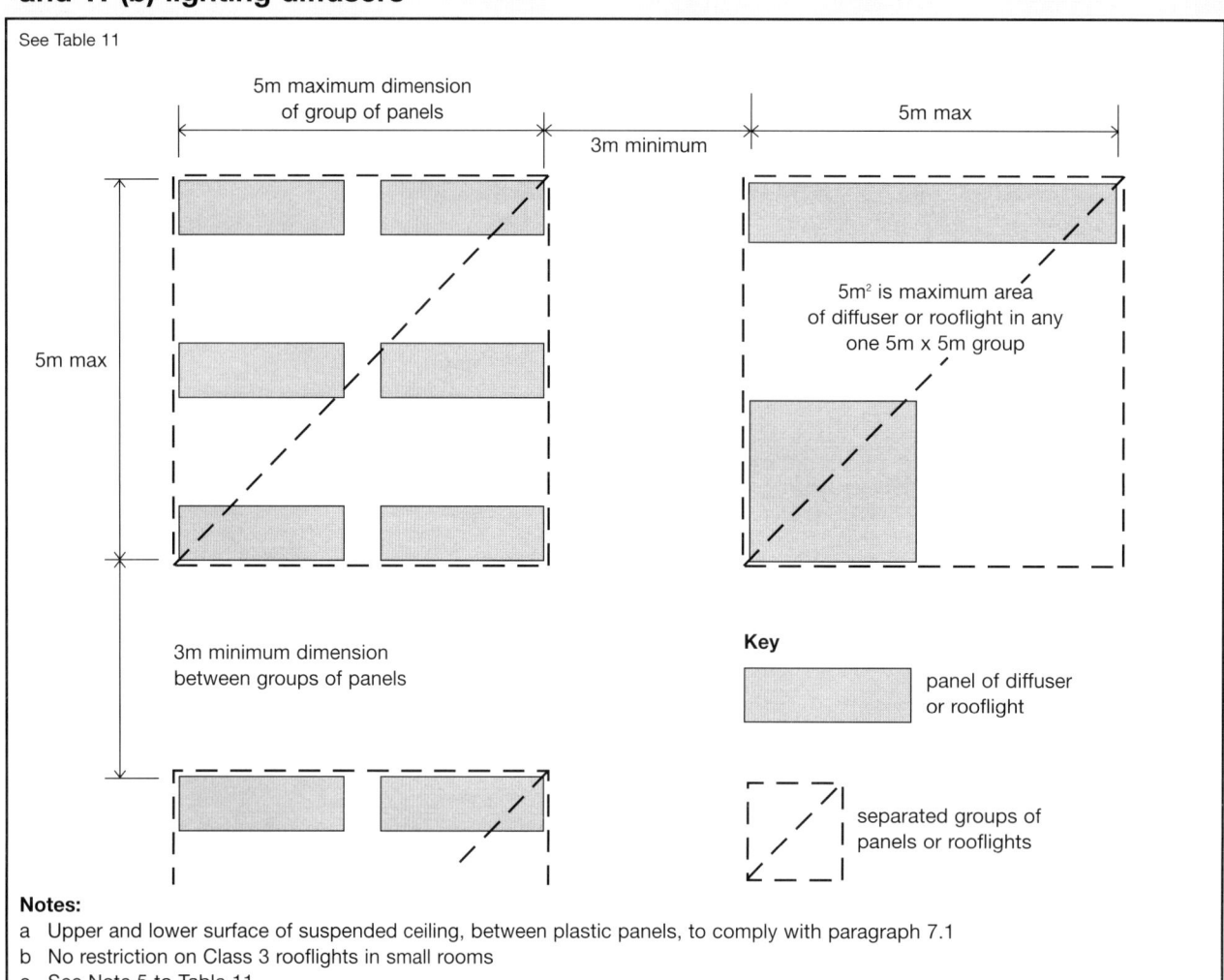

Notes:

a Upper and lower surface of suspended ceiling, between plastic panels, to comply with paragraph 7.1

b No restriction on Class 3 rooflights in small rooms

c See Note 5 to Table 11

The Requirement

This Approved Document deals with the following
Requirement from Part B of Schedule 1 to the
Building Regulations 2000.

Requirement	Limits on application
Internal fire spread (structure) **B3.-**(1) The building shall be designed and constructed so that, in the event of fire, its stability will be maintained for a reasonable period. (2) A wall common to two or more buildings shall be designed and constructed so that it adequately resists the spread of fire between those buildings. For the purposes of this sub-paragraph a house in a terrace and a semi-detached house are each to be treated as a separate building. (3) To inhibit the spread of fire within the building, it shall be sub-divided with fire-resisting construction to an extent appropriate to the size and intended use of the building. (4) The building shall be designed and constructed so that the unseen spread of fire and smoke within concealed spaces in its structure and fabric is inhibited.	 Requirement B3(3) does not apply to material alterations to any prison provided under section 33 of the Prisons Act 1952.

Guidance

Performance

In the Secretary of State's view the requirements of B3 will be met:

a. if the loadbearing elements of structure of the building are capable of withstanding the effects of fire for an appropriate period without loss of stability;

b. if the building is sub-divided by elements of fire-resisting construction into compartments;

c. if any openings in fire-separating elements (see Appendix E) are suitably protected in order to maintain the integrity of the element (ie the continuity of the fire separation); and

d. if any hidden voids in the construction are sealed and subdivided to inhibit the unseen spread of fire and products of combustion, in order to reduce the risk of structural failure, and the spread of fire, in so far as they pose a threat to the safety of people in and around the building.

The extent to which any of these measures are necessary is dependent on the use of the building, and in some cases its size, and on the location of the element of construction.

Introduction

B3.i Guidance on loadbearing elements of structure is given in Section 8. Section 9 is concerned with the subdivision of a building into compartments, and Section 10 makes provisions about concealed spaces (or cavities). Section 11 gives information on the protection of openings and on fire-stopping which relates to compartmentation and to fire spread in concealed spaces. Section 12 is concerned with special measures which apply to car parks and shopping complexes. Common to all these sections, and to other provisions of Part B, is the property of fire resistance.

Fire resistance

B3.ii The fire resistance of an element of construction is a measure of its ability to withstand the effects of fire in one or more ways, as follows:

a. resistance to collapse, ie the ability to maintain loadbearing capacity (which applies to loadbearing elements only);

b. resistance to fire penetration, ie an ability to maintain the integrity of the element;

c. resistance to the transfer of excessive heat, ie an ability to provide insulation from high temperatures.

B3.iii "Elements of structure" is the term applied to the main structural loadbearing elements, such as structural frames, floors and loadbearing walls. Compartment walls are treated as elements of structure although they are not necessarily loadbearing. Roofs, unless they serve the function of a floor, are not treated as elements of structure. External walls such as curtain walls or other forms of cladding which transmit only self weight and wind loads and do not transmit floor load are not regarded as loadbearing for the purposes of B3.ii(a) although they may need fire resistance to satisfy requirement B4 (see sections 13 – 14).

Loadbearing elements may or may not have a fire-separating function. Similarly, fire-separating elements may or may not be loadbearing.

Guidance elsewhere in the Approved Document concerning fire resistance

B3.iv There is guidance in Sections 2 – 6 concerning the use of fire-resisting construction to protect means of escape. There is guidance in Section 13 about fire resistance of external walls to restrict the spread of fire between buildings. There is guidance in Section 18 about fire resistance in the construction of firefighting shafts. Appendix A gives information on methods of test and performance for elements of construction. Appendix B gives information on fire doors. Appendix C gives information on methods of measurement. Appendix D gives information on purpose group classification. Appendix E gives definitions.

Section 8

LOADBEARING ELEMENTS OF STRUCTURE

Introduction

8.1 Premature failure of the structure can be prevented by provisions for loadbearing elements of structure to have a minimum standard of fire resistance, in terms of resistance to collapse or failure of loadbearing capacity. The purpose in providing the structure with fire resistance is threefold, namely:

a. to minimise the risk to the occupants, some of whom may have to remain in the building for some time while evacuation proceeds if the building is a large one;

b. to reduce the risk to fire fighters, who may be engaged on search or rescue operations;

c. to reduce the danger to people in the vicinity of the building, who might be hurt by falling debris or as a result of the impact of the collapsing structure on other buildings.

Fire resistance standard

8.2 Structural frames, beams, columns, loadbearing walls (internal and external), floor structures and gallery structures, should have at least the fire resistance given in Appendix A, Table A1.

Application of the fire resistance standards for loadbearing elements

8.3 The measures set out in Appendix A include provisions to ensure that where one element of structure supports or gives stability to another element of structure, the supporting element has no less fire resistance than the other element (see notes to Table A2). The measures also provide for elements of structure that are common to more than one building or compartment, to be constructed to the standard of the greater of the relevant provisions. Special provisions about fire resistance of elements of structure in single storey buildings are also given, and there are concessions in respect of fire resistance of elements of structure in basements where at least one side of the basement is open at ground level.

Exclusions from the provisions for elements of structure

8.4 The following are excluded from the definition of element of structure for the purposes of these provisions:

a. structure that only supports a roof, unless:

   i. the roof performs the function of a floor, such as for parking vehicles, or as a means of escape (see Sections 2 – 6), or

   ii. the structure is essential for the stability of an external wall which needs to have fire resistance;

b. the lowest floor of the building;

c. a platform floor; and

d. a loading gallery, fly gallery, stage grid, lighting bridge, or any gallery provided for similar purposes or for maintenance and repair (see definition of "Element of structure" in Appendix E).

Additional guidance

8.5 Guidance in other sections of this Approved Document may also apply if a loadbearing wall is:

a. a compartment wall (this includes a wall common to two buildings), (see Section 9);

b. a wall between a house and a domestic garage, (see Section 9, paragraph 9.14);

c. a wall enclosing a place of special fire hazard (see Section 9, paragraph 9.12);

d. protecting a means of escape, (see Sections 2 – 6);

e. an external wall, (see Sections 13 & 14); or

f. enclosing a firefighting shaft, (see Section 18).

8.6 If a floor is also a compartment floor, see Section 9.

Floors in domestic loft conversions

8.7 In altering an existing two storey single family dwellinghouse to provide additional storeys, the provisions in this Approved Document are for the floor(s), both old and new, to have the full 30 minute standard of fire resistance shown in Appendix A, Table A1. However provided that the following conditions are satisfied, namely:

a. only one storey is being added;

b. the new storey contains no more than 2 habitable rooms; and

c. the total area of the new storey does not amount to more than 50m²;

then the existing first floor construction may be accepted if it has at least a modified 30 minute standard of fire resistance, in those places where:

d. the floor separates only rooms (and not circulation spaces); provided that

e. the particular provisions for loft conversions in Section 2, paragraphs 2.17-2.26, are met.

Notes:

1. The 'modified 30 minute' standard satisfies the test criteria for the full 30 minutes in respect of loadbearing capacity, but allows reduced performances for integrity and insulation (see Appendix A, Table A1, item 3(a)).

2. Sub-paragraph (d) above means that the floor needs a full 30 minutes standard where it forms part of the enclosures to the circulation space between the loft conversion and the final exit.

Raised storage areas

8.8 Raised free standing floors (sometimes supported by racking) are frequently erected in single storey industrial buildings. Whether the structure is considered as a gallery or is of sufficient size that it is considered as a floor forming an additional storey, the normal provisions for fire resistance of elements of structure may be onerous if applied to the raised storage area.

8.9 A structure which does not have the appropriate fire resistance given in Appendix A, Table A1 is acceptable provided the following conditions are satisfied:

a. the structure has only one tier and is used for storage purposes **only**;

b. the number of persons likely to be on the floor at any one time is low and does not include members of the public;

c. the floor is not more than 10m in either width or length and does not exceed one half of the floor area of the space in which it is situated;

d. the floor is open above and below to the room or space in which it is situated; and

e. the means of escape from the floor meets the relevant provisions in Sections 4, 5 and 6.

Notes:

1. Where the lower level is provided with an automatic detection and alarm system meeting the relevant recommendations of BS 5839: Part 1 *Fire detection and alarm systems for buildings, Code of practice for system design, installation and servicing*, then the floor size may be increased to not more than 20m in either width or length.

2. Where the building is fitted throughout with an automatic sprinkler system meeting the relevant recommendations of BS 5306: Part 2 *Fire extinguishing installations and equipment on premises, Specification for sprinkler systems*, ie the relevant occupancy rating together with the additional requirements for life safety, there are no limits on the size of the floor.

Conversion to flats

8.10 Where an existing house or other building is converted into flats there is a material change of use to which Part B of the regulations applies. Where the existing building has timber floors and these are to be retained, the relevant provisions for fire resistance may be difficult to meet.

8.11 Provided that the means of escape conform to Section 3, and are adequately protected, a 30 minute standard of fire resistance could be accepted for the elements of structure in a building having not more than 3 storeys.

Where the altered building has 4 or more storeys the full standard of fire resistance given in Appendix A would normally be necessary.

Section 9

COMPARTMENTATION

Introduction

9.1 The spread of fire within a building can be restricted by sub-dividing it into compartments separated from one another by walls and/or floors of fire-resisting construction. The object is twofold:

a. to prevent rapid fire spread which could trap occupants of the building; and

b. to reduce the chance of fires becoming large, on the basis that large fires are more dangerous, not only to occupants and fire service personnel, but to people in the vicinity of the building. Compartmentation is complementary to provisions made in Sections 2 – 6 for the protection of escape routes, and to provisions made in Sections 13-15 against the spread of fire between buildings.

9.2 The appropriate degree of sub-division depends on:

a. the use of, and fire load in, the building, which affects the potential for fires and the severity of fires, as well as the ease of evacuation;

b. the height to the floor of the top storey in the building, which is an indication of the ease of evacuation and the ability of the fire service to intervene effectively; and

c. the availability of a sprinkler system which affects the growth rate of the fire, and may suppress it altogether.

9.3 Sub-division is achieved using compartment walls and compartment floors. The circumstances in which they are needed are given in paragraphs 9.9 to 9.20.

9.4 Provisions for the construction of compartment walls and compartment floors are given in paragraphs 9.21 et seq. These construction provisions vary according to the function of the wall or floor.

Special forms of compartmentation

9.5 Special forms of compartmentation to which particular construction provisions apply, are:

a. walls common to two or more buildings, see paragraph 9.23;

b. walls dividing buildings into separated parts, see paragraph 9.24;

c. construction enclosing places of special fire hazard, see paragraph 9.12; and

d. construction protecting houses from attached or integral domestic garages, see paragraph 9.14.

Junctions

9.6 For compartmentation to be effective, there should be continuity at the junctions of the fire-resisting elements enclosing a compartment, and any openings from one compartment to another should not present a weakness.

Protected shafts

9.7 Spaces that connect compartments, such as stairways and service shafts, need to be protected to restrict fire spread between the compartments, and they are termed protected shafts. Any walls or floors bounding a protected shaft are considered to be compartment walls or floors, for the purpose of this Approved Document.

Buildings containing one or more atria

9.8 Detailed advice on all issues relating to the incorporation of atria in buildings is given in BS 5588: Part 7 *Fire precautions in the design, construction and use of buildings, Code of practice for the incorporation of atria in buildings*. However it should be noted that for the purposes of Approved Document B, the standard is relevant only where the atrium breaches any compartmentation.

Provision of compartmentation

General

9.9 Compartment walls and compartment floors should be provided in the circumstances described below, with the proviso that the lowest floor in a building does not need to be constructed as a compartment floor. Paragraphs 9.10 – 9.20 give guidance on the provision of compartmentation in different building types. Information on the construction of compartment walls and compartment floors in different circumstances is given in paragraphs 9.21 et seq. Provisions for the protection of openings in compartment walls and compartment floors are given in paragraphs 9.33 et seq.

All purpose groups

9.10 A wall common to two or more buildings should be constructed as a compartment wall.

9.11 Parts of a building that are occupied mainly for different purposes, should be separated from one another by compartment walls and/or compartment floors. This does not apply where one of the different purposes is ancillary to the other. Refer to Appendix D for guidance on whether a function should be regarded as ancillary or not.

Places of special fire hazard

9.12 Every place of special fire hazard (see Appendix E) should be enclosed with fire-resisting construction; see Table A1, item 15.

Note: Any such walls and floors are not compartment walls and compartment floors.

Houses

9.13 Every wall separating semi-detached houses, or houses in terraces, should be constructed as a compartment wall, and the houses should be considered as separate buildings.

9.14 If a domestic garage is attached to (or forms an integral part of) a house, the garage should be separated from the rest of the house, as shown in Diagram 25.

Note: The walls and floors shown in Diagram 25 are not compartment walls and compartment floors and that the 100mm difference in level between the garage floor and the door opening is to prevent any leakage of petrol vapour into the dwelling.

Diagram 25 **Separation between garage and dwellinghouse**

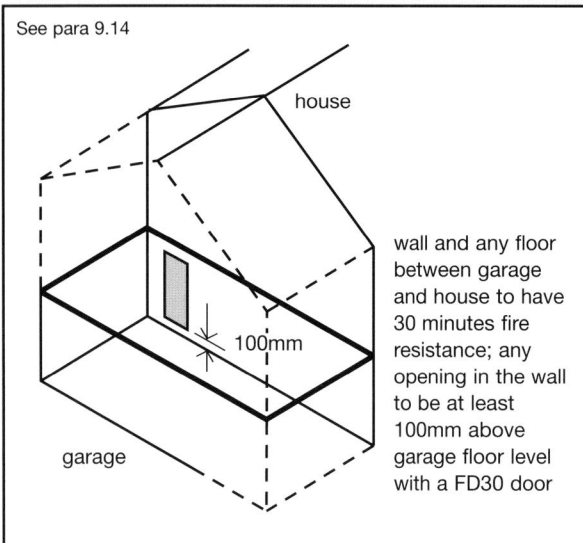

See para 9.14

house

wall and any floor between garage and house to have 30 minutes fire resistance; any opening in the wall to be at least 100mm above garage floor level with a FD30 door

100mm

garage

Flats

9.15 In buildings containing flats or maisonettes the following should be constructed as compartment walls or compartment floors:

a. every floor (unless it is within a maisonette, ie between one storey and another within one dwelling); and

b. every wall separating a flat or maisonette from any other part of the building; and

Note: "any other part of the building" does not include an external balcony/deck access.

c. every wall enclosing a refuse storage chamber.

Institutional buildings including health care

9.16 All floors should be constructed as compartment floors.

9.17 Compartments should not exceed 2000m^2 in multi-storey hospitals and 3000m^2 in single storey hospitals.

9.18 Every upper storey used for in-patient care should be divided into at least two compartments in such a way as to permit progressive horizontal evacuation of each compartment. (See Section 4, paragraph 4.31.)

Other residential buildings

9.19 All floors should be constructed as compartment floors.

Non-residential buildings

9.20 The following walls and floors should be constructed as compartment walls and compartment floors in buildings of a non-residential purpose group (ie Office, Shop & Commercial, Assembly & Recreation, Industrial, Storage or Other non-residential):

a. every wall needed to sub-divide the building to observe the size limits on compartments given in Table 12 (see Diagram 26a);

b. every floor if the building, or separated part (see paragraph 9.24) of the building, has a storey with a floor at a height of more than 30m above ground level (see Diagram 26b);

c. the floor of the ground storey if the building has one or more basements (see Diagram 26c), with the exception of small premises (see paragraph 4.1);

d. the floor of every basement storey (except the lowest floor) if the building, or separated part (see paragraph 9.24), has a basement at a depth of more than 10m below ground level (see Diagram 26d);

e. if the building forms part of a Shopping Complex, every wall and floor described in Section 5 of BS 5588: Part 10: 1991 *Fire precautions in the design, construction and use of buildings, Code of practice for shopping complexes* as needing to be constructed to the standard for a compartment wall or compartment floor; and

f. if the building comprises Shop & Commercial, Industrial or Storage premises, every wall or floor provided to divide a building into separate occupancies, (ie spaces used by different organisations whether they fall within the same Purpose Group or not).

Note: See also the provision in paragraph 6.54 for store rooms in shops to be separated from retail areas by fire-resisting construction to the standard given in Table A1.

Diagram 26 **Compartment floors: illustration of guidance in paragraph 9.20**

A. EXAMPLE OF COMPARTMENTATION IN AN UNSPRINKLERED SHOP see para 9.20(a)

None of the floors in this case would need to be compartment floors, but the 2 storeys exceeding 2000 sq.m would need to be divided into compartments not more than 2000 sq.m by compartment walls.

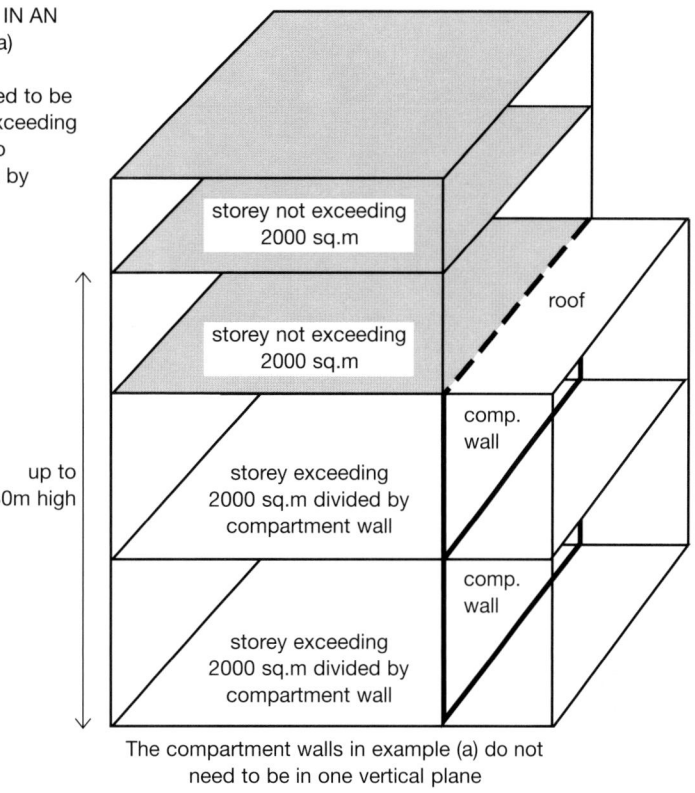

The compartment walls in example (a) do not need to be in one vertical plane

B. COMPARTMENTATION IN TALL BUILDINGS see para 9.20(b)

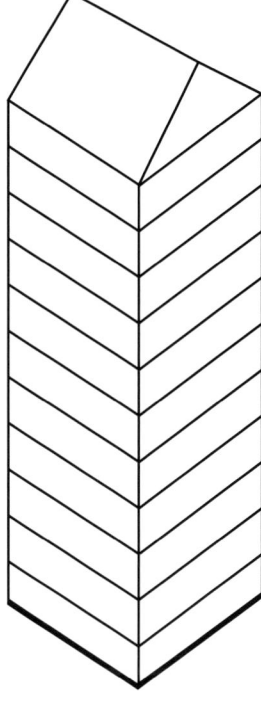

In a building over 30m in height all storeys should be separated by compartment floors. For advice on the special conditions in atrium buildings see B5588: Part 7.

C. SHALLOW BASEMENTS see para 9.20(c)

Only the floor of the ground storey need be a compartment floor if the lower basement is at a depth of not more than 10m

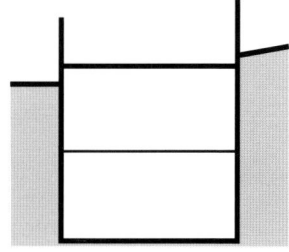

D. DEEP BASEMENTS see para 9.20(d)

All basement storeys to be separated by compartment floors if any storey is at a depth of more than 10m

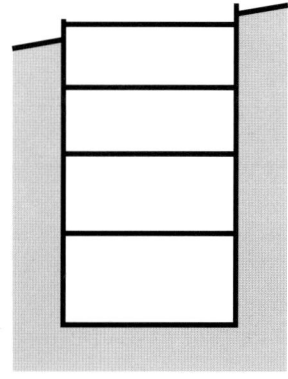

Table 12 **Maximum dimensions of building or compartment (non-residential buildings)**

Purpose Group of building or part	Height of floor of top storey above ground level (m)	Floor area of any one storey in the building or any one storey in a compartment (m²)	
		in multi-storey buildings	in single storey buildings
Office	no limit	no limit	no limit
Assembly & Recreation Shop & Commercial:			
a. schools	no limit	800	800
b. shops – not sprinklered	no limit	2000	2000
shops – sprinklered (1)	no limit	4000	no limit
c. elsewhere – not sprinklered	no limit	2000	no limit
elsewhere – sprinklered (1)	no limit	4000	no limit
Industrial (2)			
not sprinklered	not more than 18	7000	no limit
	more than 18	2000 (3)	no limit
sprinklered (1)	not more than 18	14000	no limit
	more than 18	4000 (3)	no limit
	Height of floor of top storey above ground level (m)	Maximum compartment volume (m³)	
		in multi-storey buildings	in single storey buildings
Storage (2) & Other non-residential:			
a. car park for light vehicles	no limit	no limit	no limit
b. any other building or part:			
not sprinklered	not more than 18	20000	no limit
	more than 18	4000 (3)	no limit
sprinklered (1)	not more than 18	40000	no limit
	more than 18	8000 (3)	no limit

Notes:

1. "Sprinklered" means that the building is fitted throughout with an automatic sprinkler system meeting the relevant recommendations of BS 5306: Part 2, ie the relevant occupancy rating together with the additional requirements for life safety.

2. There may be additional limitations on floor area and/or sprinkler provisions in certain industrial and storage uses under other legislation, for example in respect of storage of LPG and certain chemicals.

3. This reduced limit applies only to storeys that are more than 18m above ground level. Below this height the higher limit applies.

Construction of compartment walls and compartment floors

General

9.21 In a two storey building in the shop, commercial or industrial Purpose Groups, where the use of the upper storey is ancillary to the use of the ground storey, the ground storey may be treated as a single storey building for fire compartmentation purposes, provided that:

a. the area of the upper storey does not exceed 20% of the area of the ground storey, or 500m², whichever is less;

b. the upper storey is compartmented from the lower one; and

c. there is a means of escape from the upper storey that is independent of the routes from the lower storey.

9.22 Every compartment wall and compartment floor should:

a. form a complete barrier to fire between the compartments they separate; and

b. have the appropriate fire resistance as indicated in Appendix A, Tables A1 and A2.

Note: Timber beams, joists, purlins and rafters may be built into or carried through a masonry or concrete compartment wall if the openings for them are kept as small as practicable and then fire-stopped. If trussed rafters bridge the wall, they should be designed so that failure of any part of the truss due to a fire in one compartment will not cause failure of any part of the truss in another compartment.

Compartment walls between buildings

9.23 Compartment walls that are common to two or more buildings should run the full height of the building in a continuous vertical plane. Thus adjoining buildings should only be separated by walls, not floors.

Separated parts of buildings

9.24 Compartment walls used to form a separated part of a building (so that the separated parts can be assessed independently for the purpose of determining the appropriate standard of fire resistance) should run the full height of the building in a continuous vertical plane. The two separated parts can have different standards of fire resistance.

Other Compartment Walls

9.25 Compartment walls not described in the previous two paragraphs should run the full height of the storey in which they are situated.

9.26 Compartment walls in a top storey beneath a roof should be continued through the roof space (see definition of compartment in Appendix E).

Junction of compartment wall or compartment floor with other walls

9.27 Where a compartment wall or compartment floor meets another compartment wall, or an external wall, the junction should maintain the fire resistance of the compartmentation.

Junction of compartment wall with roof

9.28 A compartment wall should be taken up to meet the underside of the roof covering or deck, with fire-stopping where necessary at the wall/roof junction to maintain the continuity of fire resistance. The compartment wall should also be continued across any eaves cavity (see paragraph 9.22a).

9.29 If a fire penetrates a roof near a compartment wall there is a risk that it will spread over the roof to the adjoining compartment. To reduce this risk, and subject to 9.30 below, a zone of the roof 1500mm wide on either side of the wall should have a covering of designation AA, AB or AC (see Appendix A, paragraph 6) on a substrate or deck of a material of limited combustibility, as set out in Diagram 28a.

Note: Double skinned insulated roof sheeting should incorporate a band of material of limited combustibility.

Diagram 27 **Compartment walls and compartment floors with reference to relevant paragraphs in Section 9**

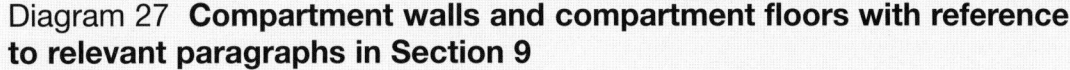

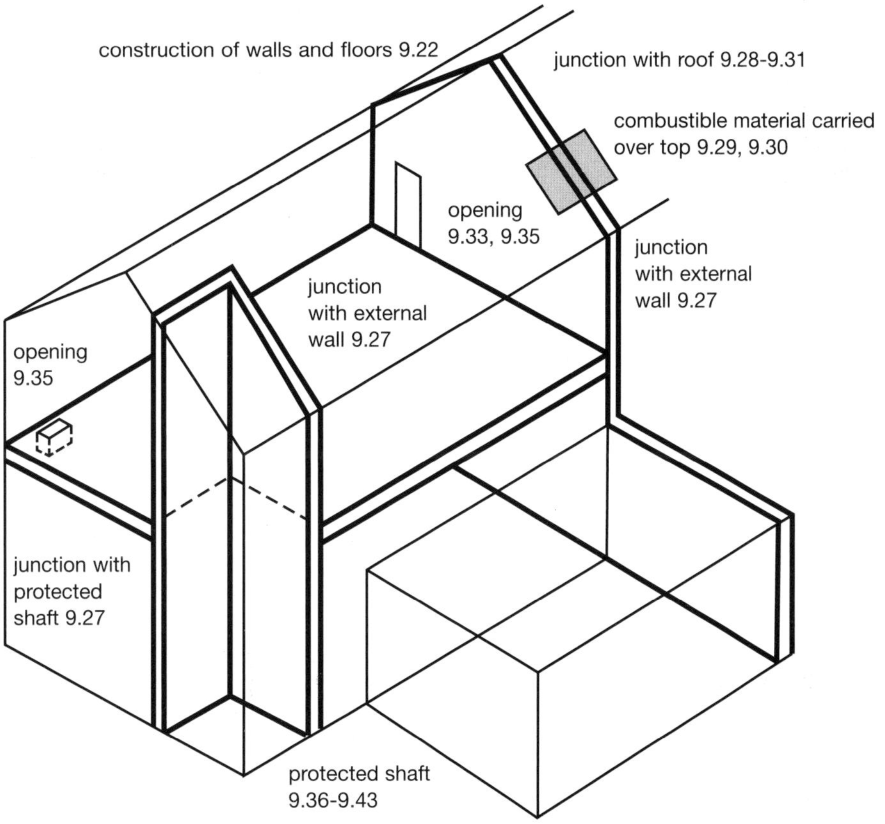

construction of walls and floors 9.22

junction with roof 9.28-9.31

combustible material carried over top 9.29, 9.30

opening 9.33, 9.35

junction with external wall 9.27

junction with external wall 9.27

opening 9.35

junction with protected shaft 9.27

protected shaft 9.36-9.43

9.30 In buildings not more than 15m high, of the purpose groups listed below, combustible boarding used as a substrate to the roof covering, wood wool slabs, or timber tiling battens, may be carried over the compartment wall provided that they are fully bedded in mortar or other suitable material over the width of the wall (see Diagram 28b). This applies to: Dwellinghouses, buildings or compartments in Residential use (other than Institutional), Office buildings, Assembly and recreation buildings.

9.31 As an alternative to 9.29 or 9.30 the compartment wall may be extended up through the roof for a height of at least 375mm above the top surface of the adjoining roof covering (see Diagram 28c).

Compartment construction in hospitals

9.32 Compartment walls and floors in hospitals designed on the basis of Firecode (see B1.xx) should be constructed of materials of limited combustibility if they have fire resistance of 60 minutes or more (unless the building is fitted throughout with a suitable sprinkler system – see Firecode).

Openings in compartmentation

Openings in compartment walls separating buildings or occupancies

9.33 Any openings in a compartment wall which is common to two or more buildings, or between different occupancies in the same building, should be limited to those for:

a. a door which is needed to provide a means of escape in case of fire and which has the same fire resistance as that required for the wall (see Appendix B, Table B1) and is fitted in accordance with the provisions of Appendix B; and

b. the passage of a pipe which meets the provisions in Section 11.

Doors

9.34 Information on fire doors will be found in Appendix B.

Openings in other compartment walls or in compartment floors

9.35 Openings in compartment walls (other than those described in paragraph 9.33) or compartment floors should be limited to those for:

a. doors which have the appropriate fire resistance given in Appendix B, Table B1, and are fitted in accordance with the provisions of Appendix B; and

b. the passage of pipes, ventilation ducts, chimneys, appliance ventilation ducts or ducts encasing one or more flue pipes, which meet the provisions in Section 11; and

c. refuse chutes of non-combustible construction; and

d. atria designed in accordance with BS 5588: Part 7; and

e. protected shafts which meet the relevant provisions below.

Protected shafts

9.36 Any stairway or other shaft passing directly from one compartment to another should be enclosed in a protected shaft so as to delay or prevent the spread of fire between compartments.

There are additional provisions in Sections 2 – 6 for protected shafts that are protected stairways, and in Section 18 if the stairway also serves as a firefighting stair.

Diagram 28 **Junction of compartment wall with roof**

See paras 9.28-9.31

a. ANY BUILDING OR COMPARTMENT

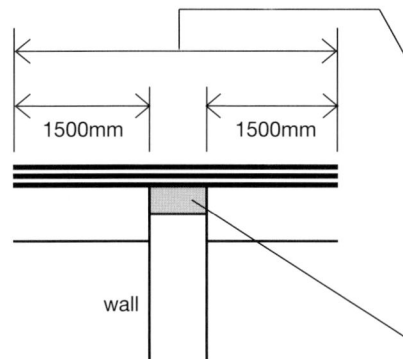

1500mm 1500mm

wall

Roof covering over this distance to be designated AA, AB or AC on deck of material of limited combustibility. Roof covering and deck could be composite structure, eg profiled steel cladding.

Double skinned insulated roof sheeting should incorporate a band of material of limited combustibility at least 300mm wide centred over the wall.

If roof support members pass through the wall, fire protection to these members for a distance of 1500mm on either side of the wall may be needed to delay distortion at the junction (see Note to para 9.22).

Resilient fire stopping to be carried up to underside of roof covering.

b. DWELLINGHOUSE AND BUILDING OR COMPARTMENT IN RESIDENTIAL (NOT INSTITUTIONAL), OFFICE OR ASSEMBLY USE and not more than 15m high

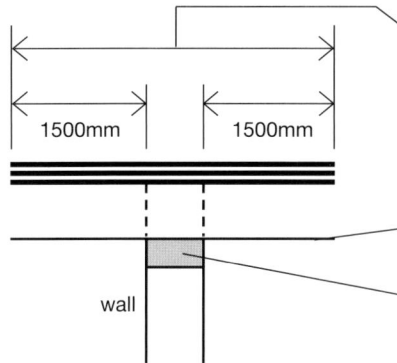

1500mm 1500mm

wall

Roof covering to be designated AA, AB or AC for at least this distance.

Boarding (used as a substrate), wood wool slabs or timber tiling battens may be carried over the wall provided that they are fully bedded in mortar (or other no less suitable material) where over the wall.

Sarking felt may also be carried over the wall.

If roof support members pass through the wall, fire protection to these members for a distance of 1500mm on either side of the wall may be needed to delay distortion at the junction (see Note to para 9.22)

Fire stopping to be carried up to underside of roof covering, boarding or slab.

c. ANY BUILDING OR COMPARTMENT

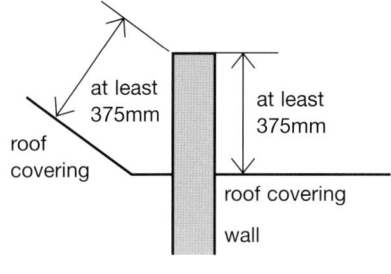

at least 375mm

at least 375mm

roof covering

roof covering

wall

Diagram 29 **Protected shafts**

See paras 9.36-9.38

Protected shafts provide for the movement of people (eg stairs, lifts), or for passage of goods, air or services such as pipes or cables between different compartments. The elements enclosing the shaft (unless formed by adjacent external walls) are compartment walls and floors. The diagram shows three common examples which illustrate the principles.

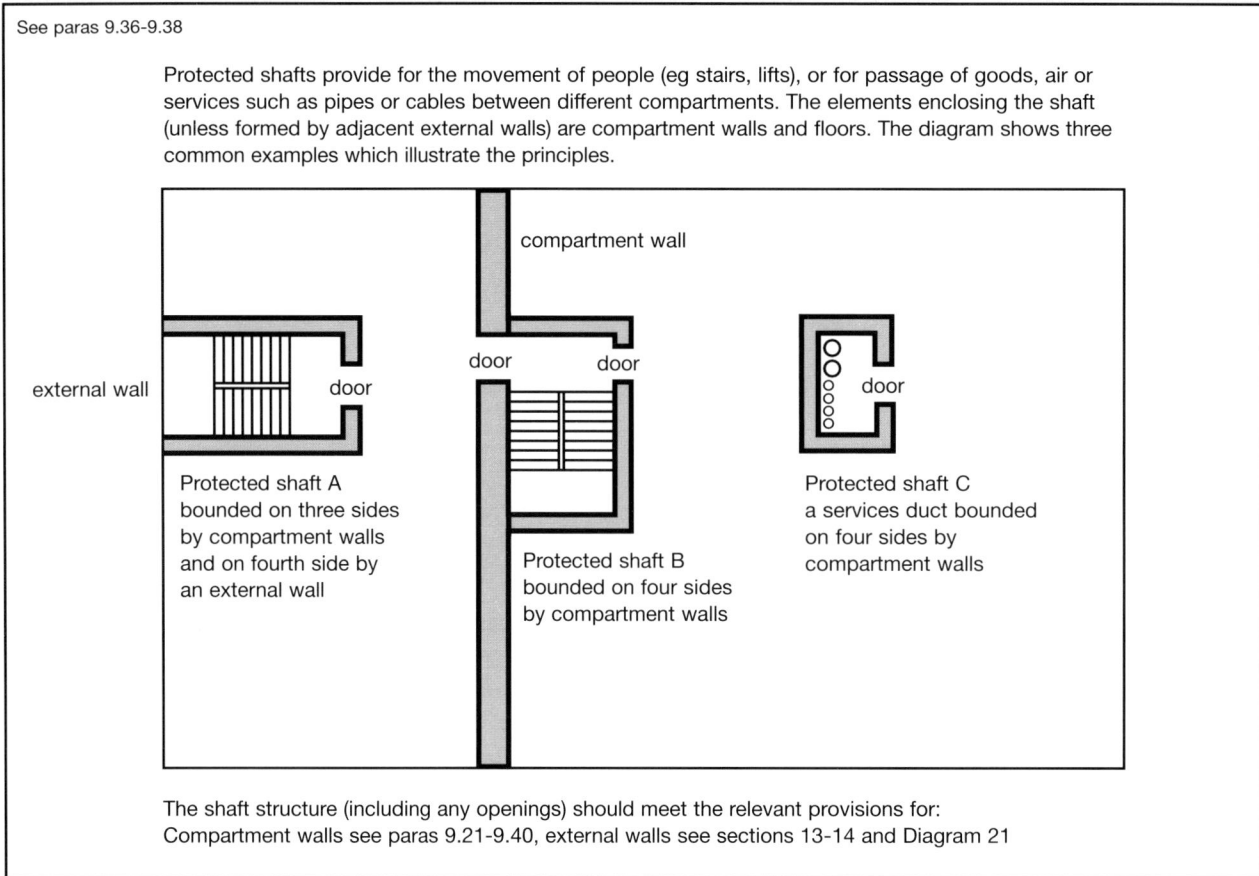

The shaft structure (including any openings) should meet the relevant provisions for:
Compartment walls see paras 9.21-9.40, external walls see sections 13-14 and Diagram 21

Uses for protected shafts

9.37 The uses of protected shafts should be restricted to stairs, lifts, escalators, chutes, ducts, and pipes. Sanitary accommodation and washrooms may be included in protected shafts.

Construction of protected shafts

9.38 The construction enclosing a protected shaft (see Diagram 29) should:

a. form a complete barrier to fire between the different compartments which the shaft connects;

b. have the appropriate fire resistance given in Appendix A, Table A1, except for uninsulated glazed screens which meet the provisions of paragraph 9.39; and

c. satisfy the provisions about their ventilation and the treatment of openings in paragraphs 9.42 et seq.

Uninsulated glazed screens to protected shafts

9.39 If the conditions given below and described in Diagram 30 are satisfied, an uninsulated glazed screen may be incorporated in the enclosure to a protected shaft between a stair and a lobby or corridor which is entered from the stair. The conditions to be satisfied are:

a. the standard of fire resistance for the stair enclosure is not more than 60 minutes; and

b. the protected shaft is not a firefighting shaft (if it is, refer to BS 5588: Part 5 *Fire precautions in the design, construction and use of buildings, Code of practice for firefighting stairs and lifts*, clauses on construction);

c. the glazed screen:

    i. has at least 30 minutes fire resistance, in terms of integrity, and

    ii. meets the guidance in Appendix A, Table A4, on the limits on areas of uninsulated glazing;

d. the lobby or corridor is enclosed to at least a 30 minute standard.

9.40 Where the measures in Diagram 30 to protect the lobby or corridor, are not provided, the enclosing walls should comply with Appendix A, Table A1 (item 8c) and the doors with the guidance in Appendix A, Table A4.

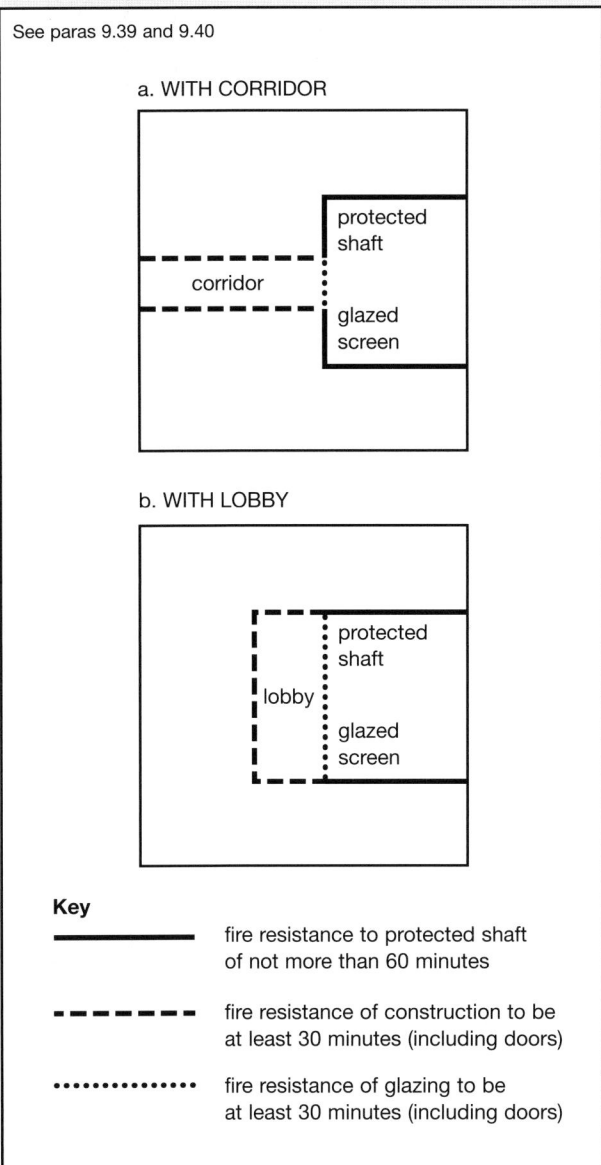

Diagram 30 **Uninsulated glazed screen separating protected shaft from lobby or corridor**

See paras 9.39 and 9.40

a. WITH CORRIDOR

protected shaft

corridor

glazed screen

b. WITH LOBBY

protected shaft

lobby

glazed screen

Key

────── fire resistance to protected shaft of not more than 60 minutes

▬ ▬ ▬ ▬ fire resistance of construction to be at least 30 minutes (including doors)

•••••••••••• fire resistance of glazing to be at least 30 minutes (including doors)

Pipes for oil or gas, and ventilating ducts, in protected shafts

9.41 If a protected shaft contains a stair and/or a lift, it should not also contain a pipe conveying oil (other than in the mechanism of a hydraulic lift) or contain a ventilating duct (other than a duct provided for the purposes for pressurizing the stairway to keep it smoke free or a duct provided solely for ventilating the stairway).

Any pipe carrying natural gas or LPG in such a shaft should be of screwed steel or of all welded steel construction, installed in accordance with the Pipelines Safety Regulations 1996, SI 1996 No 825, and the Gas Safety (Installation and use) Regulations 1998, SI 1998 No 2451.

Note: A pipe is not considered to be contained within a protected shaft, if the pipe is completely separated from that protected shaft by fire-resisting construction.

Ventilation of protected shafts conveying gas

9.42 A protected shaft conveying piped flammable gas should be adequately ventilated direct to the outside air by ventilation openings at high and low level in the shaft.

Any extension of the storey floor into the shaft should not compromise the free movement of air over the entire length of the shaft. Guidance on such shafts, including sizing of the ventilation openings, is given in BS 8313 *Code of practice for accommodation of building services in ducts*.

Openings into protected shafts

9.43 Generally an external wall of a protected shaft does not need to have fire resistance.

However there are some provisions for fire resistance of external walls of firefighting shafts in section 2 of BS 5588: Part 5: 1991, which is the relevant guidance called up by paragraph 18.11, and of external walls to protected stairways (which may also be protected shafts) in paragraph 6.24.

Openings in other parts of the enclosure to a protected shaft should be limited as follows.

a. Where part of the enclosure to a protected shaft is a wall common to two or more buildings, only the following openings should be made in that wall:

i. a door which is needed to provide a means of escape in case of fire and which has the same fire resistance as that required for the wall (see Appendix B, Table B1) and is fitted in accordance with the provisions of Appendix B; and

ii. the passage of a pipe which meets the provisions in Section 11.

b. Other parts of the enclosure (other than an external wall) should only have openings for:

i. doors which have the appropriate fire resistance given in Appendix B, Table B1, and are fitted in accordance with the provisions of Appendix B; and

ii. the passage of pipes which meet the provisions in Section 11; and

iii. inlets to, outlets from and openings for a ventilation duct, (if the shaft contains or serves as a ventilating duct) which meet the provisions in Section 11; and

iv. the passage of lift cables into a lift machine room (if the shaft contains a lift). If the machine room is at the bottom of the shaft, the openings should be as small as practicable.

Section 10

CONCEALED SPACES (CAVITIES)

Introduction

10.1 Concealed spaces or cavities in the construction of a building provide a ready route for smoke and flame spread. This is particularly so in the case of voids above other spaces in a building, eg above a suspended ceiling or in a roof space. As any spread is concealed, it presents a greater danger than would a more obvious weakness in the fabric of the building. Provisions are made to restrict this by interrupting cavities which could form a pathway around a barrier to fire, sub-dividing extensive cavities, and by closing the edges of openings.

Note: Cavity barriers are not appropriate for use above compartment walls (see paragraph 10.5). See also 'Limitation on requirements' on page 5 which explains the purpose of provisions made in connection with Building Regulations.

Note on cavities in rain screen cladding and the like: Cavities within an external wall are referred to in this Section, including the drained and ventilated cavities behind the outer cladding in "rainscreen" external wall construction. There are also provisions in paragraphs 13.6 and 13.7 about the construction of external walls which have a bearing on overcladding and rainscreen construction.

Provision of cavity barriers

10.2 Provisions for cavity barriers are set out in Table 13 against specified locations and purpose groups.

10.3 Table 14 lays down maximum dimensions for undivided concealed spaces.

10.4 Diagram 31 illustrates the need for cavity barriers at the intersection of fire-resisting construction and elements containing a concealed space.

Diagram 31 **Interrupting concealed spaces (cavities)**

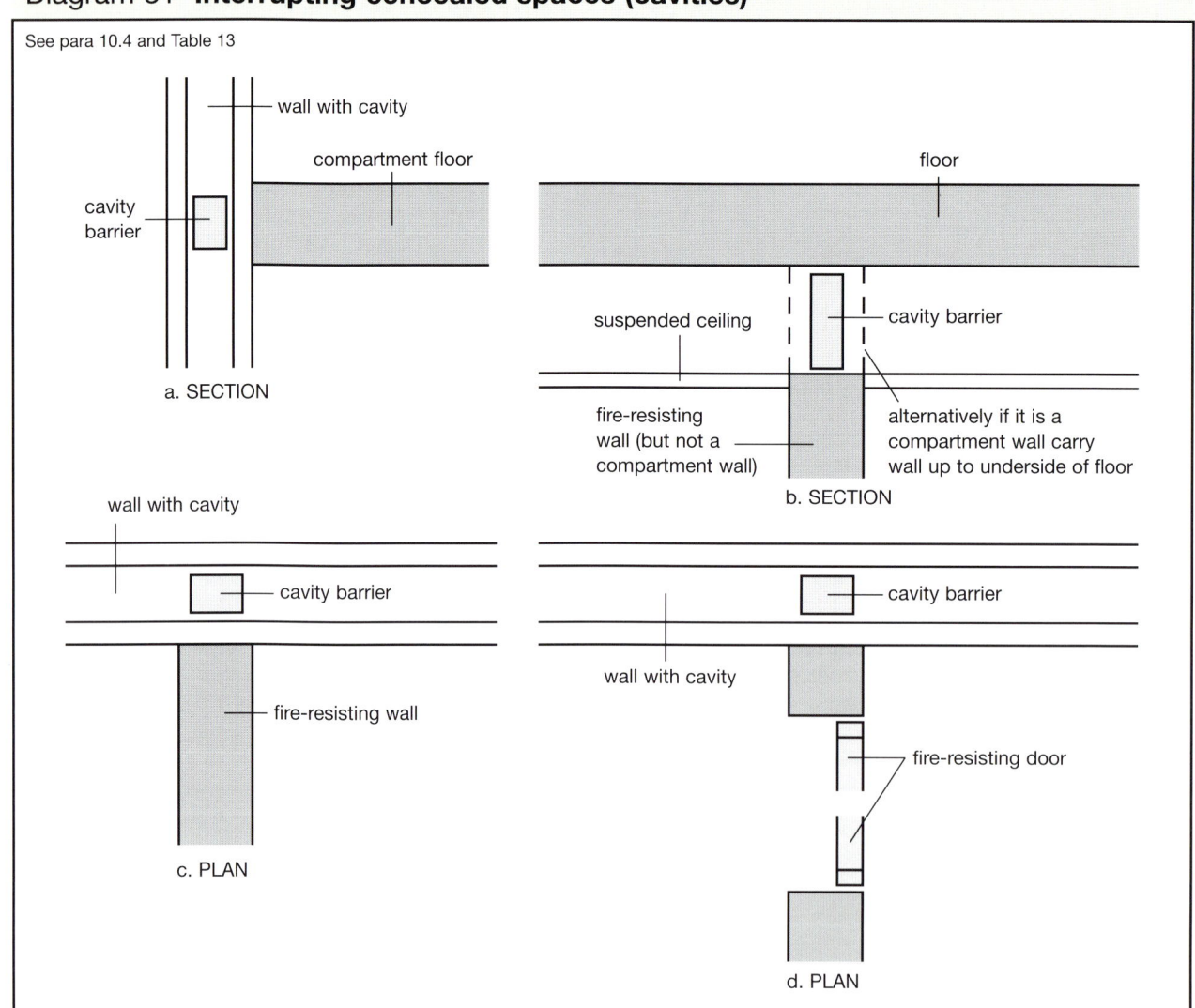

See para 10.4 and Table 13

wall with cavity

compartment floor

floor

cavity barrier

a. SECTION

suspended ceiling

cavity barrier

fire-resisting wall (but not a compartment wall)

alternatively if it is a compartment wall carry wall up to underside of floor

b. SECTION

wall with cavity

cavity barrier

fire-resisting wall

c. PLAN

cavity barrier

wall with cavity

fire-resisting door

d. PLAN

Table 13 **Provision of cavity barriers**

Cavity barriers to be provided:	Purpose group to which the provision applies (1)			
	1b & c **Dwelling** **houses**	**1a** **Flat or** **maisonette**	**2** **Other** **residential** **and** **Institutional**	**3-7** **Office, Shop &** **Commercial,** **Assembly &** **Recreation, Industrial,** **Storage & Other** **non-residential**
1. At the junction between an external cavity wall and a compartment wall that separates buildings; and at the top of such an external cavity wall. (2)	●	●	●	●
2. Above the enclosures to a protected stairway in a house with a floor more than 4.5m above ground level (see Diagram 33a). (3)	●	○	○	○
3. At the junction between an external cavity wall and every compartment floor and compartment wall. (2)	○	●	●	●
4. At the junction between a cavity wall and every compartment floor, compartment wall, or other wall or door assembly which forms a fire-resisting barrier. (2)	○	●	●	●
5. In a protected escape route, above and below any fire-resisting construction which is not carried full storey height, or (in the case of a top storey) to the underside of the roof covering. (3)	○	●	●	●
6. Where the corridor should be sub-divided to prevent fire or smoke affecting two alternative escape routes simultaneously (see paragraph 4.23 & Diagram 34a), above any such corridor enclosures which are not carried full storey height, or (in the case of the top storey) to the underside of the roof covering. (4)	○	○	●	●
7. Above any bedroom partitions which are not carried full storey height, or (in the case of the top storey) to the underside of the roof covering. (3)	○	○	●	○
8. To sub-divide any cavity (including any roof space but excluding any underfloor service void) so that the distance between cavity barriers does not exceed the dimensions given in Table 14.	○	○	●	●
9. Within the void behind the external face of rainscreen cladding at every floor level, and on the line of compartment walls abutting the external wall, of buildings which have a floor 18m or more above ground level.	○	●	●	○
10. At the edges of cavities (including around openings).	●	●	●	●

Key: ● provision applies
　　　○ provision does not apply

Notes:

1. The classification of purpose groups is set out in Appendix D, Table D1.

2. The provisions in items 1, 3 and 4 do not apply where the cavity wall complies with Diagram 32.

3. The provisions in items 2, 5 and 7 do not apply where the cavity is enclosed on the lower side by a fire-resisting ceiling (as shown in Diagram 35) which extends throughout the building, compartment or separated part.

4. The provision of item 6 does not apply where the storey is sub-divided by fire-resisting construction carried full storey height and passing through the line of sub-division of the corridor (see Diagram 34b), or where the cavity is enclosed on the lower side as described in Note 3.

Table 14   Maximum dimensions of cavities in non-domestic buildings (Purpose Groups 2-7)

Location of cavity	Class of surface/product exposed in cavity (excluding the surface of any pipe, cable or conduit, or any insulation to any pipe)		Maximum dimensions in any direction (m)
	National class	European class	
Between roof and a ceiling	Any	Any	20
Any other cavity	Class 0 or Class 1	Class A1or Class A2-s3, d2 or Class B-s3, d2 or Class C-s3, d2	20
	Not Class 0 or Class 1	Not any of the above classes	10

Notes:

1   Exceptions to these provisions are given in paragraphs 10.11-10.13.

2   The National classifications do not automatically equate with the equivalent classifications in the European column, therefore products cannot typically assume a European class unless they have been tested accordingly.

3   When a classification includes "s3, d2", this means that there is no limit set for smoke production and/or flaming droplets/particles.

Diagram 32   Cavity walls excluded from provisions for cavity barriers

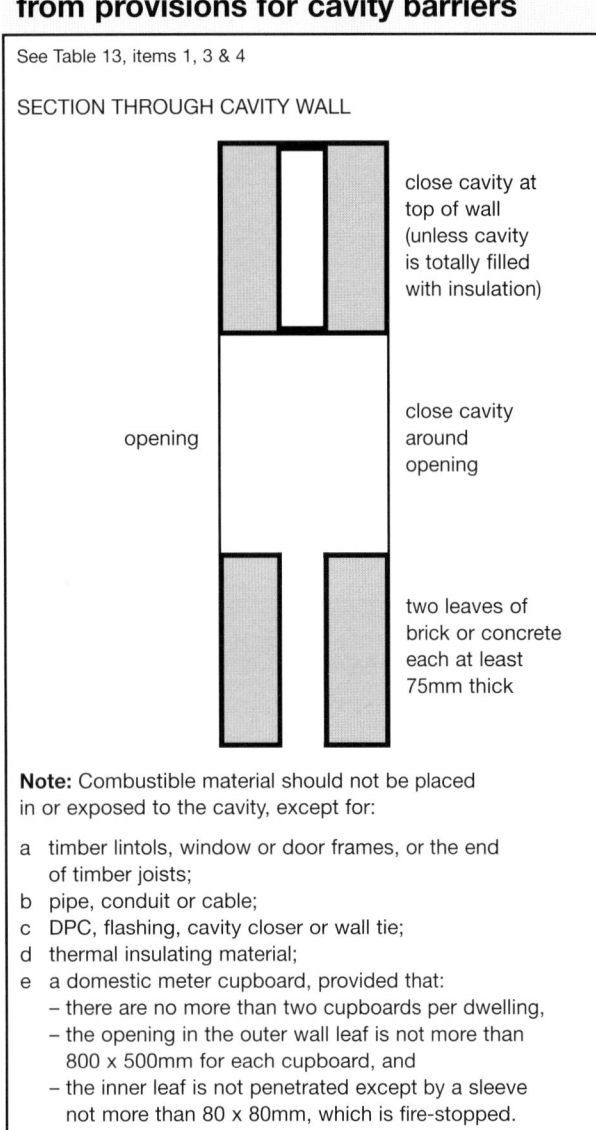

See Table 13, items 1, 3 & 4

SECTION THROUGH CAVITY WALL

close cavity at top of wall (unless cavity is totally filled with insulation)

opening

close cavity around opening

two leaves of brick or concrete each at least 75mm thick

Note: Combustible material should not be placed in or exposed to the cavity, except for:

a   timber lintols, window or door frames, or the end of timber joists;

b   pipe, conduit or cable;

c   DPC, flashing, cavity closer or wall tie;

d   thermal insulating material;

e   a domestic meter cupboard, provided that:
  - there are no more than two cupboards per dwelling,
  - the opening in the outer wall leaf is not more than 800 x 500mm for each cupboard, and
  - the inner leaf is not penetrated except by a sleeve not more than 80 x 80mm, which is fire-stopped.

Diagram 33   Alternative arrangements in roof space over protected stairway in a house with a floor more than 4.5m above ground level

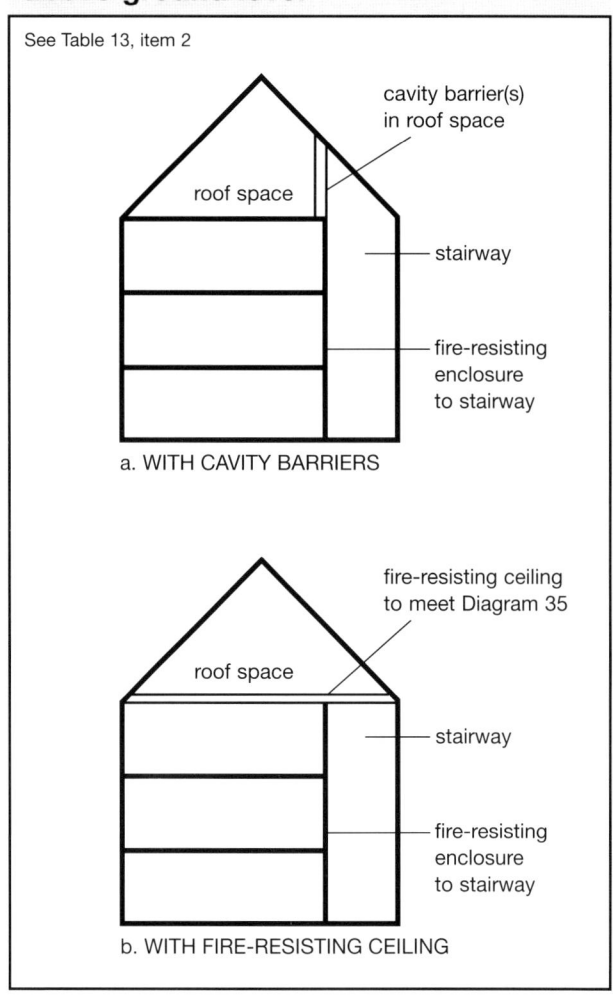

See Table 13, item 2

cavity barrier(s) in roof space

roof space

stairway

fire-resisting enclosure to stairway

a. WITH CAVITY BARRIERS

fire-resisting ceiling to meet Diagram 35

roof space

stairway

fire-resisting enclosure to stairway

b. WITH FIRE-RESISTING CEILING

Diagram 34 **Corridor enclosure alternatives**

See Table 13, item 6 & Note 4, and paragraph 4.23

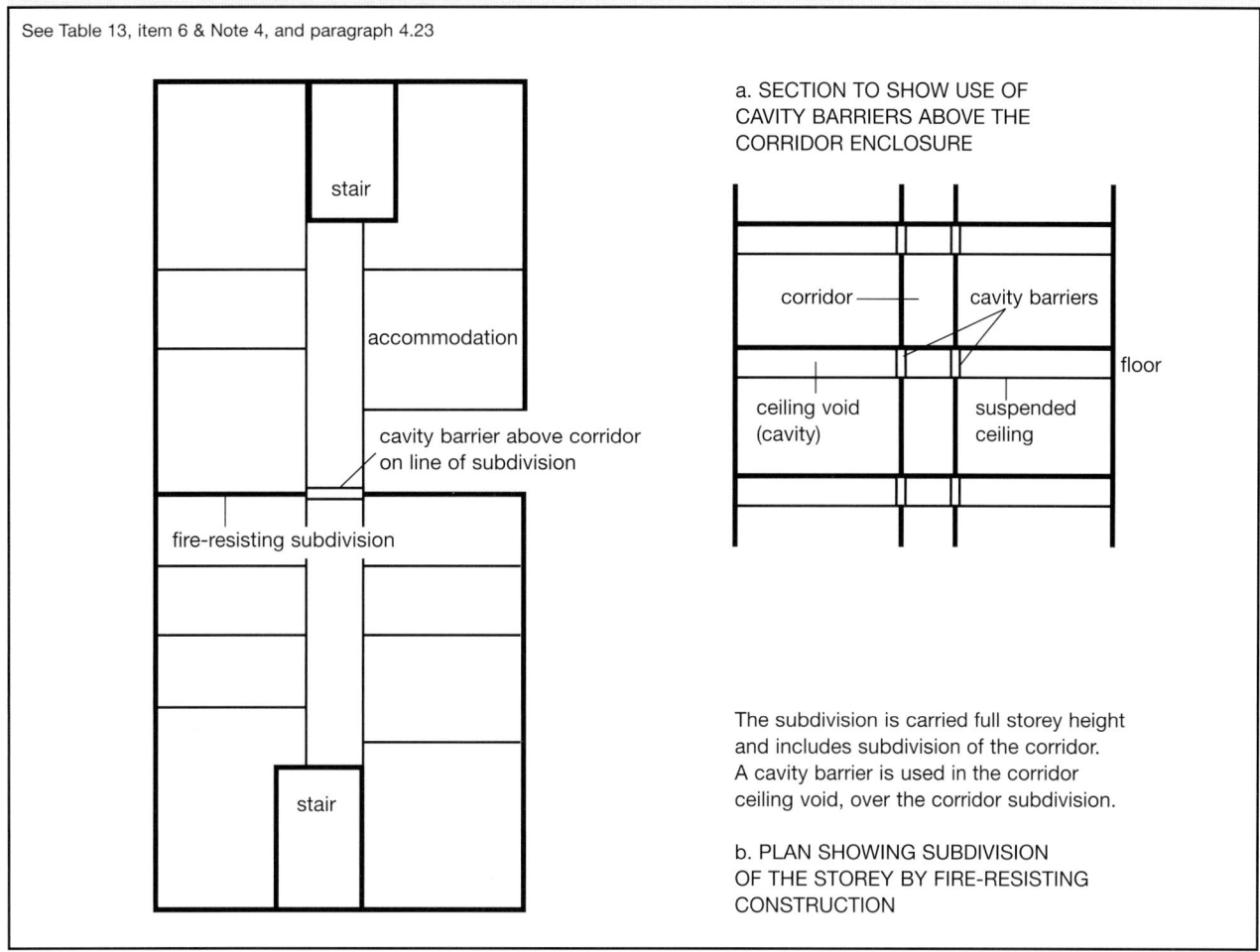

a. SECTION TO SHOW USE OF CAVITY BARRIERS ABOVE THE CORRIDOR ENCLOSURE

The subdivision is carried full storey height and includes subdivision of the corridor. A cavity barrier is used in the corridor ceiling void, over the corridor subdivision.

b. PLAN SHOWING SUBDIVISION OF THE STOREY BY FIRE-RESISTING CONSTRUCTION

10.5 As compartment walls should be carried up full storey height to a compartment floor or to the roof as appropriate, see paragraphs 9.22-9.25, it is not appropriate to complete a line of compartmentation by fitting cavity barriers above them. Therefore it is important to continue the compartment wall through the cavity to maintain the standard of fire resistance.

Construction and fixings for cavity barriers

10.6 Every cavity barrier should be constructed to provide at least 30 minutes fire resistance (see Appendix A, Table A1, item 16). However, cavity barriers in a stud wall or partition may be formed of:

a. steel at least 0.5mm thick; or

b. timber at least 38mm thick; or

c. polythene sleeved mineral wool, or mineral wool slab, in either case under compression when installed in the cavity; or

d. calcium silicate, cement based or gypsum based boards at least 12.0mm thick.

Note: Cavity barriers provided around openings (see Table 13, item 10) may be formed by the window or door frame.

10.7 A cavity barrier may be formed by any construction provided for another purpose if it meets the provisions for cavity barriers.

10.8 Cavity barriers should be tightly fitted to rigid construction and mechanically fixed in position wherever possible. Where this is not possible (for example, in the case of a junction with slates, tiles, corrugated sheeting or similar materials) the junction should be fire-stopped. Provisions for fire-stopping are set out in Section 11.

10.9 Cavity barriers should also be fixed so that their performance is unlikely to be made ineffective by:

a. movement of the building due to subsidence, shrinkage or temperature change, and movement of the external envelope due to wind; and

b. collapse in a fire of any services penetrating them; and

c. failure in a fire of their fixings (but see note below); and

d. failure in a fire of any material or construction which they abut. (For example, if a suspended ceiling is continued over the top of a fire-resisting wall or partition, and direct connection is made between the ceiling and the cavity barrier above the line of the wall or partition, premature failure of the cavity barrier can occur when the ceiling collapses. However, this does not arise if the ceiling is designed to provide fire protection of 30 minutes or more.)

Note: Where cavity barriers are provided in roof spaces, the roof members to which they are fitted are not expected to have any fire resistance.

Maximum dimensions of concealed spaces

10.10 With the exceptions given in paragraphs 10.11 to 10.13, extensive concealed spaces should be sub-divided to comply with the dimensions in Table 14.

10.11 The provisions in Table 14 do not apply to any cavity described below:

a. in a wall which should be fire-resisting only because it is loadbearing;

b. in a masonry or concrete external cavity wall shown in Diagram 32;

c. in any floor or roof cavity above a fire-resisting ceiling, as shown in Diagram 35 and which extends throughout the building or compartment, subject to a 30m limit on the extent of the cavity;

d. below a floor next to the ground or oversite concrete, if the cavity is less than 1000mm in height or if the cavity is not normally accessible by persons, unless there are openings in the floor such that it is possible for combustibles to accumulate in the cavity (in which case cavity barriers should be provided, and access should be provided to the cavity for cleaning);

e. within an underfloor service void;

f. formed behind the external skin in rain screen external wall construction, or by overcladding an existing masonry (or concrete) external wall, or an existing concrete roof, provided that the cavity does not contain combustible insulation, and the provisions of Table 13 item 9 are observed; and

g. between double-skinned corrugated or profiled insulated roof sheeting, if the sheeting is a material of limited combustibility and both surfaces of the insulating layer have a surface spread of flame of at least Class 0 or 1 (National class) or Class C-s3, d2 or better (European class) (see Appendix A) and make contact with the inner and outer skins of cladding (see Diagram 36).

   Note: When a classification includes "s3, d2", this means that there is no limit set for smoke production and/or flaming droplets/particles.

10.12 Where any room under a ceiling cavity exceeds the dimensions given in Table 14, cavity barriers need only be provided on the line of the enclosing walls/partitions of that room, subject to:

a. the cavity barriers being no more than 40m apart; and

b. the surface of the material/product exposed in the cavity being Class 0 or Class 1 (National class) or Class C-s3, d2 or better (European class).

Note: When a classification includes "s3, d2", this means that there is no limit set for smoke production and/or flaming droplets/particles.

10.13 Where the concealed space is over an undivided area which exceeds 40m (this may be in both directions on plan) there is no limit to the size of the cavity if:

a. the room and the cavity together are compartmented from the rest of the building;

b. an automatic fire detection and alarm system meeting the relevant recommendations of BS 5839: Part 1 *Fire detection and alarm systems for buildings, Code of practice for system design, installation and servicing* is fitted in the building (however, detectors are not required in the cavity);

c. if the cavity is used as a plenum, the recommendations about recirculating air distribution systems in BS 5588: Part 9 *Fire precautions in the design, construction and use of buildings, Code of practice for ventilation and air conditioning ductwork* are followed;

d. the surface of the material/product used in the construction of the ceiling and which is exposed in the cavity is Class 0 (National class) or Class B-s3, d2 or better (European class) and the supports and fixings in the cavity are of non-combustible construction;

   Note: When a classification includes "s3, d2", this means that there is no limit set for smoke production and/or flaming droplets/particles.

e. the flame spread rating of any pipe insulation system is Class 1;

f. any electrical wiring in the void is laid in metal trays, or in metal conduit; and

g. any other materials in the cavity are of limited combustibility.

Openings in cavity barriers

10.14 Any openings in a cavity barrier (except barriers identified in Table 13, item 7) should be limited to those for:

a. doors which have at least 30 minutes fire resistance (see Appendix B, Table B1, item 8(a)) and are fitted in accordance with the provisions of Appendix B;

b. the passage of pipes which meet the provisions in Section 11;

c. the passage of cables or conduits containing one or more cables;

d. openings fitted with a suitably mounted automatic fire damper; and

e. ducts which (unless they are fire-resisting) are fitted with a suitably mounted automatic fire damper where they pass through the cavity barrier.

Diagram 35 **Fire-resisting ceiling below concealed space**

See paragraph 10.11(c) and Table 13 Note 3

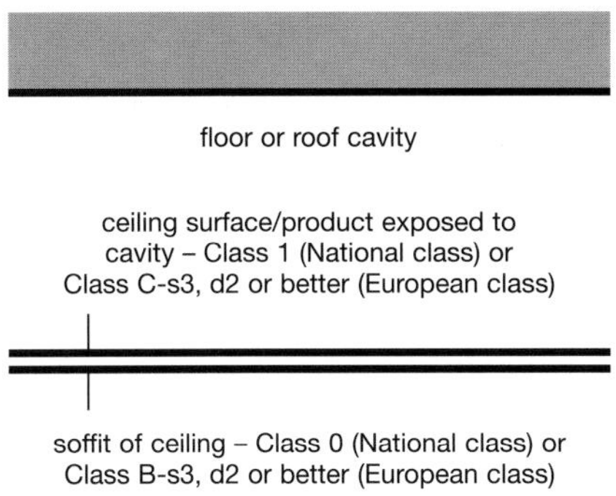

floor or roof cavity

ceiling surface/product exposed to
cavity – Class 1 (National class) or
Class C-s3, d2 or better (European class)

soffit of ceiling – Class 0 (National class) or
Class B-s3, d2 or better (European class)

Notes:
1 The ceiling should:
  a. have at least 30 minutes fire resistance;
  b. be imperforate, except for an opening described in paragraph 10.14;
  c. extend throughout the building or compartment; and
  d. not be easily demountable.
2 The National classifications do not automatically equate with the equivalent classifications in the European column, therefore products cannot typically assume a European class unless they have been tested accordingly.
3 When a classification includes "s3, d2", this means that there is no limit set for smoke production and/or flaming droplets/particles.

Diagram 36 **Provisions for cavity barriers in double-skinned insulated roof sheeting**

See para 10.11(g)

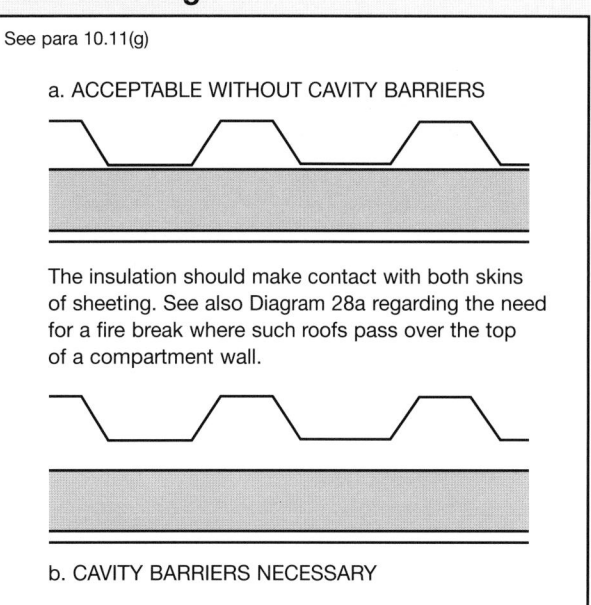

a. ACCEPTABLE WITHOUT CAVITY BARRIERS

The insulation should make contact with both skins of sheeting. See also Diagram 28a regarding the need for a fire break where such roofs pass over the top of a compartment wall.

b. CAVITY BARRIERS NECESSARY

Section 11

PROTECTION OF OPENINGS AND FIRE STOPPING

Introduction

11.1 Sections 9 and 10 make provisions for fire separating elements, and set out the circumstances in which there may be openings in them. This section deals with the protection of openings in such elements.

11.2 If a fire separating element is to be effective, then every joint, or imperfection of fit, or opening to allow services to pass through the element, should be adequately protected by sealing or fire-stopping so that the fire resistance of the element is not impaired.

11.3 The measures in this section are intended to delay the passage of fire. They generally have the additional benefit of retarding smoke spread, but the test specified in Appendix A for integrity does not stipulate criteria for the passage of smoke as such.

11.4 Detailed guidance on door openings and fire doors is given in Appendix B.

Openings for pipes

11.5 Pipes which pass through a compartment wall or compartment floor (unless the pipe is in a protected shaft), or through a cavity barrier, should meet the appropriate provisions in alternatives A, B or C below.

Alternative A: Proprietary seals (any pipe diameter)

11.6 Provide a proprietary sealing system which has been shown by test to maintain the fire resistance of the wall, floor or cavity barrier.

Alternative B: Pipes with a restricted diameter

11.7 Where a proprietary sealing system is not used, fire-stopping may be used around the pipe, keeping the opening as small as possible. The nominal internal diameter of the pipe should not be more than the relevant dimension given in Table 15.

Table 15 Maximum nominal internal diameter of pipes passing through a compartment wall/floor (see para 11.5 et seq)

Situation	Pipe material and maximum nominal internal diameter (mm)		
	(a) Non-combustible material (1)	(b) Lead, aluminium, aluminium alloy, uPVC (2), fibre cement	(c) Any other material
1. Structure (but not a wall separating buildings) enclosing a protected shaft which is not a stairway or a lift shaft	160	110	40
2. Wall separating dwelling houses, or compartment wall or compartment floor between flats	160	160 (stack pipe) (3) 110 (branch pipe) (3)	40
3. Any other situation	160	40	40

Notes:

1. Any non-combustible material (such as cast iron, copper or steel) which if exposed to a temperature of 800°C, will not soften or fracture to the extent that flame or hot gas will pass through the wall of the pipe.

2. uPVC pipes complying with BS 4514 and uPVC pipes complying with BS 5255.

3. These diameters are only in relation to pipes forming part of an above-ground drainage system and enclosed as shown in Diagram 38. In other cases the maximum diameters against situation 3 apply.

11.8 The diameters given in Table 15 for pipes of specification (b) used in situation (2) assume that the pipes are part of an above-ground drainage system and are enclosed as shown in Diagram 38. If they are not, the smaller diameter given in situation (3) should be used.

Alternative C: Sleeving

11.9 A pipe of lead, aluminium, aluminium alloy, fibre-cement or uPVC, with a maximum nominal internal diameter of 160mm, may be used with a sleeving of non-combustible pipe as shown in Diagram 37. The specification for non-combustible and uPVC pipes is given in the notes to Table 15.

Diagram 37 **Pipes penetrating structure**

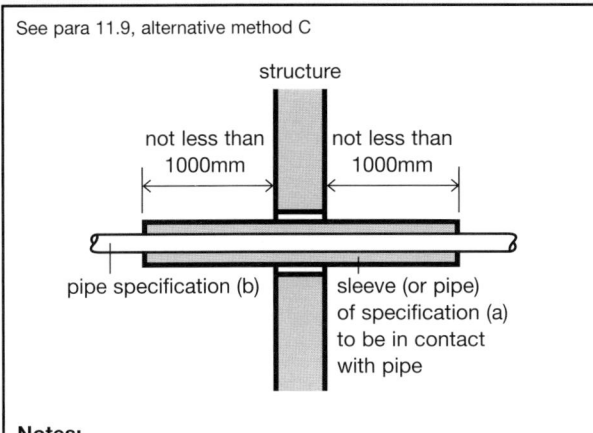

See para 11.9, alternative method C

structure

not less than 1000mm   not less than 1000mm

pipe specification (b)   sleeve (or pipe) of specification (a) to be in contact with pipe

Notes:
1   Make the opening in the structure as small as possible and provide fire-stopping between pipe and structure.
2   See Table 15 for materials specification.

Ventilating ducts

11.10 BS 5588: Part 9 *Fire precautions in the design, construction and use of buildings, Code of practice for ventilation and air conditioning ductwork* sets out alternative ways in which the integrity of compartments may be maintained where ventilation and air conditioning ducts penetrate fire separating elements. The alternatives are equally acceptable, and the recommendations of that code should be followed where air handling ducts pass from one compartment to another.

Flues, etc.

11.11 If a flue, or duct containing flues or appliance ventilation duct(s), passes through a compartment wall or compartment floor, or is built into a compartment wall, each wall of the flue or duct should have a fire resistance of at least half that of the wall or floor in order to prevent the by-passing of the compartmentation (see Diagram 39).

Fire-stopping

11.12 In addition to any other provisions in this document for fire-stopping:

a.   joints between fire separating elements should be fire-stopped; and

b.   all openings for pipes, ducts, conduits or cables to pass through any part of a fire separating element should be:

    i.   kept as few in number as possible, and

    ii.   kept as small as practicable, and

    iii.   fire-stopped (which in the case of a pipe or duct, should allow thermal movement).

11.13 To prevent displacement, materials used for fire-stopping should be reinforced with (or supported by) materials of limited combustibility in the following circumstances:

a.   in all cases where the unsupported span is greater than 100mm; and

b.   in any other case where non-rigid materials are used (unless they have been shown to be satisfactory by test).

11.14 Proprietary fire-stopping and sealing systems, (including those designed for service penetrations) which have been shown by test to maintain the fire resistance of the wall or other element, are available and may be used.

Other fire-stopping materials include:

• cement mortar,

• gypsum based plaster,

• cement or gypsum based vermiculite/perlite mixes,

• glass fibre, crushed rock, blast furnace slag or ceramic based products (with or without resin binders), and

• intumescent mastics.

These may be used in situations appropriate to the particular material. Not all of them will be suitable in every situation.

Diagram 38 **Enclosure for drainage or water supply pipes**

See para 11.8 and Table 15

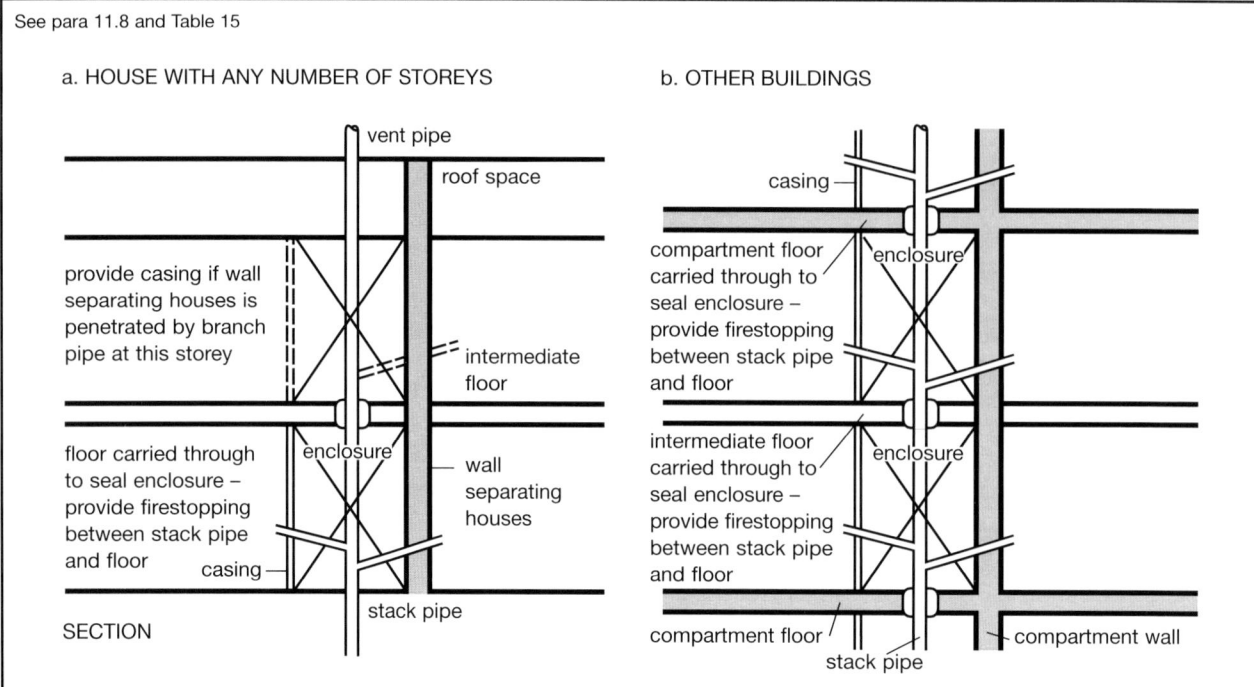

a. HOUSE WITH ANY NUMBER OF STOREYS

b. OTHER BUILDINGS

Notes:
1 The enclosure should:
   a be bounded by a compartment wall or floor, an outside wall, an intermediate floor, or a casing (see specification at 2 below), and
   b have internal surfaces (except framing members) of Class 0 (National class) or Class B-s3, d2 or better (European class)
      Note: When a classification includes "s3, d2", this means that there is no limit set for smoke production and/or flaming droplets/particles), and
   c not have an access panel which opens into a circulation space or bedroom, and
   d be used only for drainage, or water supply, or vent pipes for a drainage system
2 The casing should:
   a be imperforate except for an opening for a pipe or an access panel, and
   b not be of sheet metal, and
   c have (including any access panel) not less than 30 minutes fire resistance
3 The opening for a pipe, either in the structure or the casing, should be as small as possible and fire-stopped around the pipe.

Diagram 39 **Flues penetrating compartment walls or floors**
(note that there is guidance in Approved Document J concerning hearths adjacent to compartment walls)

See para 11.11

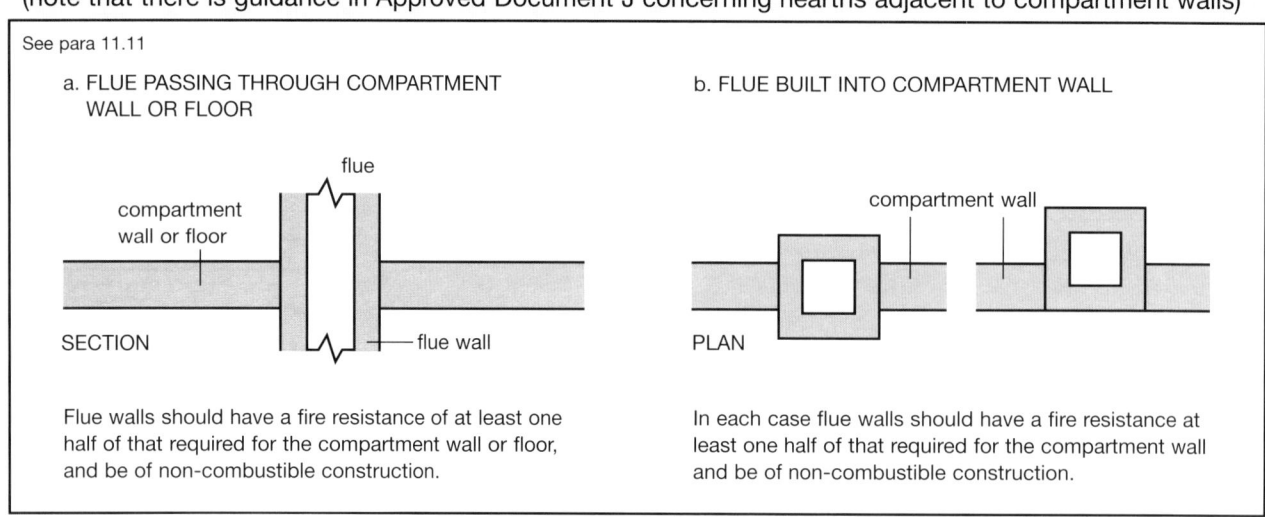

a. FLUE PASSING THROUGH COMPARTMENT
   WALL OR FLOOR

b. FLUE BUILT INTO COMPARTMENT WALL

Flue walls should have a fire resistance of at least one half of that required for the compartment wall or floor, and be of non-combustible construction.

In each case flue walls should have a fire resistance at least one half of that required for the compartment wall and be of non-combustible construction.

Section 12

SPECIAL PROVISIONS FOR CAR PARKS AND SHOPPING COMPLEXES

Introduction

12.1 This section describes additional considerations which apply to the design and construction of car parks and shopping complexes.

Car Parks

General principles

12.2 Buildings or parts of buildings used as parking for cars and other light vehicles are unlike other buildings in certain respects which merit some departures from the usual measures to restrict fire spread within buildings.

a. The fire load is well defined and not particularly high.

b. Where the car park is well ventilated, there is a low probability of fire spread from one storey to another. Ventilation is the important factor, and as heat and smoke cannot be dissipated so readily from a car park that is not open-sided fewer concessions are made. The guidance in paragraphs 12.4 to 12.7 is concerned with three ventilation methods: open sided (high level of natural ventilation), natural ventilation and mechanical ventilation.

Note: Because of the above, car parks are not normally expected to be fitted with sprinklers.

Provisions common to all car park buildings

12.3 The relevant provisions of the guidance on requirements B1 and B5 will apply, but in addition all materials used in the construction of the building, compartment or separated part should be non-combustible, except for:

a. any surface finish applied:

   i.   to a floor or roof of the car park, or

   ii. within any adjoining building, compartment or separated part to the structure enclosing the car park, if the finish meets all relevant aspects of the guidance on requirements B2 and B4;

b. any fire door;

c. any attendant's kiosk not exceeding 15m² in area; and

d. any shop mobility facility.

Open sided car parks

12.4 If the building, or separated part containing the car park, complies with the following provisions (in addition to those in paragraph 12.3) it may be regarded as an open sided car park for the purposes of fire resistance assessment in Appendix A, Table A2. The provisions are that:

a. there should not be any basement storeys;

b. each storey should be naturally ventilated by permanent openings at each car parking level, having an aggregate vent area not less than 1/20th of the floor area at that level, of which at least half (1/40th) should be equally provided between two opposing walls;

c. if the building is also used for any other purpose, the part forming the car park is a separated part.

Car parks which are not open sided

12.5 Where car parks do not have the standard of ventilation set out in 12.4(b), they are not regarded as open-sided and a different standard of fire resistance is necessary (the relevant provisions are given in Appendix A, Table A2).

Such car parks still require some ventilation, which may be by natural or mechanical means, as described in 12.6 or 12.7 below. The provisions of 12.3 apply to all car park buildings, whatever standard of ventilation is provided.

Natural ventilation

12.6 Where car parks that are not open-sided are provided with some, more limited, natural ventilation, each storey should be ventilated by permanent openings at each car parking level, having an aggregate vent area not less than 1/40th of the floor area at that level, of which at least half (1/80th) should be equally provided between two opposing walls. Smoke vents at ceiling level may be used as an alternative to the provision of permanent openings in the walls. They should have an aggregate area of permanent opening totalling not less than 1/40th of the floor area and be arranged to provide a through draught. (See Approved Document F *Ventilation* for additional guidance on normal ventilation of car parks.)

Mechanical ventilation

12.7 In most basement car parks, and in enclosed car parks, it may not be possible to obtain the minimum standard of natural ventilation openings set out in paragraph 12.6 above. In such cases a system of mechanical ventilation should be provided as follows:

a. the system should be independent of any other ventilating system (other than any system providing normal ventilation to the car park) and be designed to operate at 10 air changes per hour in a fire condition. (See Approved Document F *Ventilation* for guidance on normal ventilation of car parks);

b. the system should be designed to run in two parts, each part capable of extracting 50% of the rates set out in (a) above, and designed so that each part may operate singly or simultaneously;

c. each part of the system should have an independent power supply which would operate in the event of failure of the main supply;

d. extract points should be arranged so that 50% of the outlets are at high level, and 50% at low level: and

e. the fans should be rated to run at 300°C for a minimum of 60 mins, and the ductwork and fixings should be constructed of materials having a melting point not less than 800°C.

For further information on equipment for removing hot smoke refer to BS 7346: Part 2 *Components for smoke and heat control systems, Specification for powered smoke and heat exhaust ventilators*.

An alternative method of providing smoke ventilation from enclosed car parks is given in the BRE Report *Design methodologies for smoke and heat exhaust ventilation* (BR 368, 1999).

Shopping Complexes

12.8 Whilst the provisions in this document about shops should generally be capable of application in cases where a shop is contained in a single separate building, the provisions may not be appropriate where a shop forms part of a complex. These may include covered malls providing access to a number of shops and common servicing areas. In particular the provisions about maximum compartment size may be difficult to meet bearing in mind that it would generally not be practical to compartment a shop from a mall serving it. To a lesser extent the provisions about fire resistance, walls separating shop units, surfaces and boundary distances may pose problems.

12.9 To ensure a satisfactory standard of fire safety in shopping complexes, alternative measures and additional compensatory features to those set out in this document are appropriate. Such features are set out in sections 5 and 6 of BS 5588: Part 10: 1991 *Fire precautions in the design, construction and use of buildings, Code of practice for shopping complexes*, and the relevant recommendations of those sections should be followed.

The Requirement

This Approved Document deals with the following
Requirement from Part B of Schedule 1 to the
Building Regulations 2000.

Requirement	Limits on application
External fire spread **B4.**-(1) The external walls of the building shall adequately resist the spread of fire over the walls and from one building to another, having regard to the height, use and position of the building. (2) The roof of the building shall adequately resist the spread of fire over the roof and from one building to another, having regard to the use and position of the building.	

Guidance

Performance

In the Secretary of State's view the requirements of B4 will be met:

a. if the external walls are constructed so that the risk of ignition from an external source, and the spread of fire over their surfaces, is restricted by making provision for them to have low rates of heat release;

b. if the amount of unprotected area in the side of the building is restricted so as to limit the amount of thermal radiation that can pass through the wall, taking the distance between the wall and the boundary into account;

c. if the roof is constructed so that the risk of spread of flame and/or fire penetration from an external fire source is restricted.

In each case so as to limit the risk of a fire spreading from the building to a building beyond the boundary, or vice versa.

The extent to which this is necessary is dependent on the use of the building, its distance from the boundary and, in some cases, its height.

Introduction

External walls

B4.i The construction of external walls and the separation between buildings to prevent external fire spread are closely related.

The chances of fire spreading across an open space between buildings, and the consequences if it does, depend on:

a. the size and intensity of the fire in the building concerned;

b. the distance between the buildings;

c. the fire protection given by their facing sides; and

d. the risk presented to people in the other building(s).

B4.ii Provisions are made in Section 13 for the fire resistance of external walls and to limit the susceptibility of the external surface of walls to ignition and to fire spread.

B4.iii Provisions are made in Section 14 to limit the extent of openings and other unprotected areas in external walls in order to reduce the risk of fire spread by radiation.

Roofs

B4.iv Provisions are made in Section 15 for reducing the risk of fire spread between roofs and over the surfaces of roofs.

Section 13

CONSTRUCTION OF EXTERNAL WALLS

Introduction

13.1 Provisions are made in this section for the external walls of the building to have sufficient fire resistance to prevent fire spread across the relevant boundary. The provisions are closely linked with those for space separation in Section 14 (following) which sets out limits on the amount of unprotected area of wall. As the limits depend on the distance of the wall from the relevant boundary, it is possible for some or all of the walls to have no fire resistance, except for any parts which are loadbearing (see paragraph B3.ii).

External walls are elements of structure and the relevant period of fire resistance (specified in Appendix A) depends on the use, height and size of the building concerned. If the wall is 1000mm or more from the relevant boundary, a reduced standard of fire resistance is accepted in most cases and the wall only needs fire resistance from the inside.

13.2 Provisions are also made to restrict the combustibility of external walls of buildings that are less than 1000mm from the relevant boundary and, irrespective of boundary distance, the external walls of high buildings and those of the Assembly and Recreation Purpose Group. This is in order to reduce the surface's susceptibility to ignition from an external source, and to reduce the danger from fire spread up the external face of the building.

In the guidance to Requirement B3, provisions are made in Section 8 for internal and external loadbearing walls to maintain their loadbearing function in the event of fire.

Fire resistance standard

13.3 The external walls of the building should have the appropriate fire resistance given in Appendix A, Table A1, unless they form an unprotected area under the provisions of Section 14.

Portal frames

13.4 Portal frames are often used in single storey industrial and commercial buildings where there may be no need for fire resistance of the structure (Requirement B3). However where a portal framed building is near a relevant boundary, the external wall near the boundary may need fire resistance to restrict the spread of fire between buildings.

It is generally accepted that a portal frame acts as a single structural element because of the moment-resisting connections used, especially at the column/rafter joints. Thus in cases where the external wall of the building cannot be wholly unprotected, the rafter members of the frame, as well as the column members, may need to be fire protected.

Following an investigation of the behaviour of steel portal frames in fire, it is considered technically and economically feasible to design the foundation and its connection to the portal frame so that it would transmit the overturning moment caused by the collapse, in a fire, of unprotected rafters, purlins and some roof cladding while allowing the external wall to continue to perform its structural function. The design method for this is set out in *Fire and steel construction: The behaviour of steel portal frames in boundary conditions*, 1990 (2nd edition), which is available from the Steel Construction Institute, Silwood Park, Ascot, Berks SL5 7QN. This publication offers guidance on many aspects of portal frames, including multi-storey types.

Notes:

1. The recommendations in the SCI publication for designing the foundation to resist overturning need not be followed if the building is fitted with a sprinkler system meeting the relevant recommendations of BS 5306: Part 2 *Fire extinguishing installations and equipment on premises, Specification for sprinkler systems*, ie the relevant occupancy rating together with the additional requirements for life safety.

2. Normally, portal frames of reinforced concrete can support external walls requiring a similar degree of fire resistance without specific provision at the base to resist overturning.

3. Existing buildings may have been designed to the following guidance which is also acceptable:

   a. the column members are fixed rigidly to a base of sufficient size and depth to resist overturning;

   b. there is brick, block or concrete protection to the columns up to a protected ring beam providing lateral support; and

   c. there is some form of roof venting to give early heat release. (The roof venting could be, for example, pvc rooflights covering some 10 per cent of the floor area and evenly spaced over the floor area.)

External surfaces

13.5 The external surfaces of walls should meet the provisions in Diagram 40. However, the total amount of combustible material may be limited in practice by the provisions for space separation in Section 14 (see paragraph 14.7 et seq). Where a mixed use building includes Assembly and Recreation Purpose Group accommodation, the external surfaces of walls should meet the provisions in Diagram 40c.

Note: One alternative to meeting the provisions in Diagram 40 could be BRE Fire Note 9 *Assessing the fire performance of external cladding systems: a test method* (BRE, 1999).

13.6 In the case of the outer cladding of a wall of 'rainscreen' construction (with a drained and ventilated cavity), the surface of the outer cladding which faces the cavity should also meet the provisions of Diagram 40.

External wall construction

13.7 The external envelope of a building should not provide a medium for fire spread if it is likely to be a risk to health or safety. The use of combustible materials for cladding framework, or of combustible thermal insulation as an overcladding or in ventilated cavities, may present such a risk in tall buildings, even though the provisions for external surfaces in Diagram 40 may have been satisfied.

In a building with a storey 18m or more above ground level, insulation material used in ventilated cavities in the external wall construction should be of limited combustibility (see Appendix A). This restriction does not apply to masonry cavity wall construction which complies with Diagram 32 in Section 10.

Advice on the use of thermal insulation material is given in the BRE Report *Fire performance of external thermal insulation for walls of multi-storey buildings* (BR 135, 1988).

Diagram 40 **Provisions for external surfaces of walls**

See paras 13.5 and 13.6

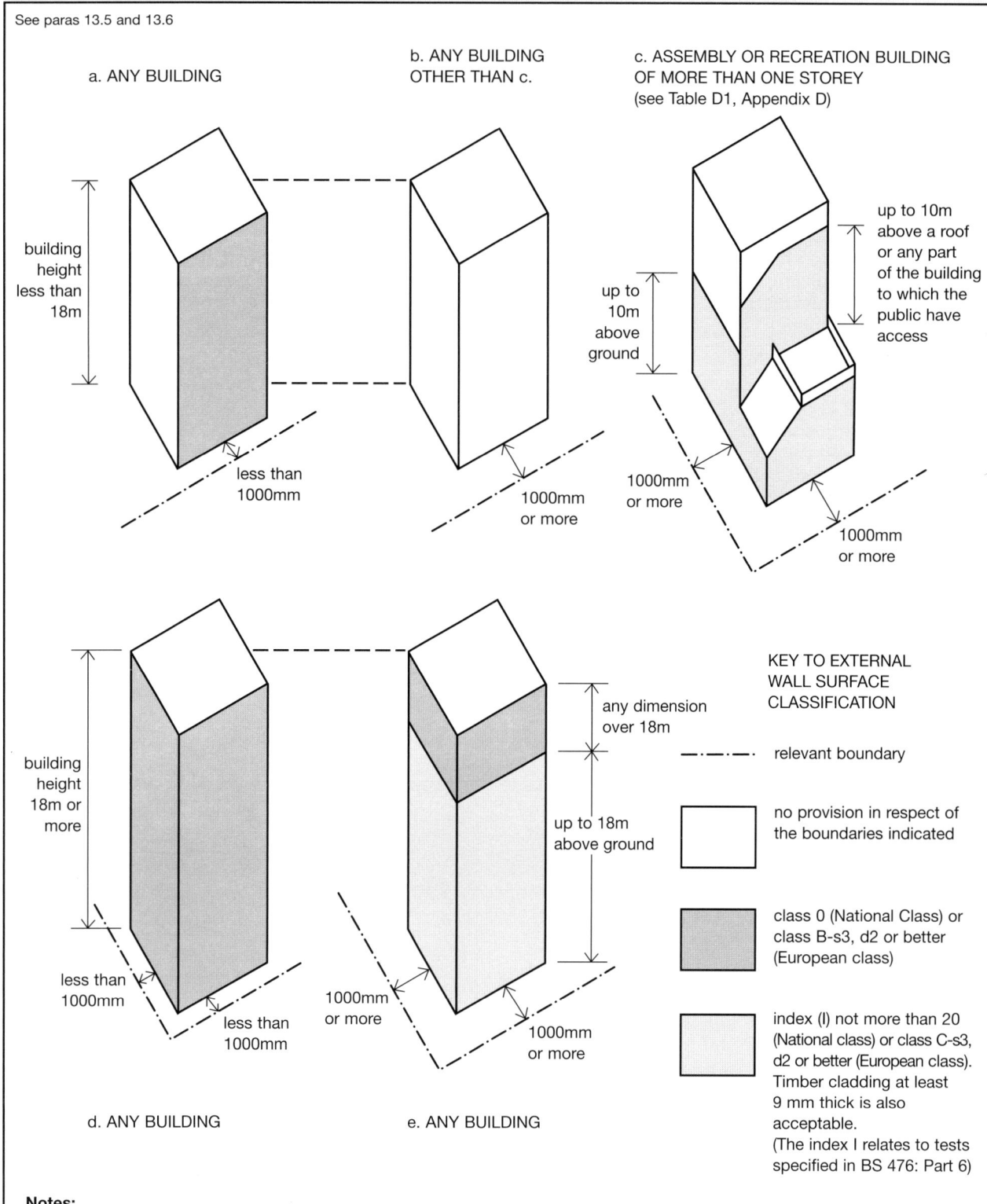

a. ANY BUILDING

b. ANY BUILDING OTHER THAN c.

c. ASSEMBLY OR RECREATION BUILDING OF MORE THAN ONE STOREY (see Table D1, Appendix D)

building height less than 18m

up to 10m above ground

up to 10m above a roof or any part of the building to which the public have access

less than 1000mm

1000mm or more

1000mm or more

1000mm or more

building height 18m or more

any dimension over 18m

up to 18m above ground

less than 1000mm

less than 1000mm

1000mm or more

1000mm or more

d. ANY BUILDING

e. ANY BUILDING

KEY TO EXTERNAL WALL SURFACE CLASSIFICATION

— · — · — relevant boundary

no provision in respect of the boundaries indicated

class 0 (National Class) or class B-s3, d2 or better (European class)

index (I) not more than 20 (National class) or class C-s3, d2 or better (European class). Timber cladding at least 9 mm thick is also acceptable. (The index I relates to tests specified in BS 476: Part 6)

Notes:

1. The National classifications do not automatically equate with the equivalent European classifications, therefore products cannot typically assume a European class unless they have been tested accordingly.
2. When a classification includes "s3, d2", this means that there is no limit set for smoke production and/or flaming droplets/particles.

Section 14

SPACE SEPARATION

Introduction

14.1 The provisions in this Section are based on a number of assumptions, and whilst some of these may differ from the circumstances of a particular case, together they enable a reasonable standard of space separation to be specified. The provisions limit the extent of unprotected areas in the sides of a building (such as openings and areas with a combustible surface) which will not give adequate protection against the external spread of fire from one building to another.

A roof is not subject to the provisions in this Section unless it is pitched at an angle greater than 70° to the horizontal (see definition for 'external wall' in Appendix E). Similarly, vertical parts of a pitched roof such as dormer windows (which taken in isolation might be regarded as a wall), would not need to meet the following provisions unless the slope of the roof exceeds 70°. It is a matter of judgement whether a continuous run of dormer windows occupying most of a steeply pitched roof should be treated as a wall rather than a roof.

14.2 The assumptions are:

a. that the size of a fire will depend on the compartmentation of the building, so that a fire may involve a complete compartment, but will not spread to other compartments;

b. that the intensity of the fire is related to the use of the building (ie purpose group), but that it can be moderated by a sprinkler system;

c. that Residential, and Assembly and Recreation, Purpose Groups represent a greater life risk than other uses;

d. that apart from Residential, and Assembly and Recreation Purpose Groups, the spread of fire between buildings on the same site represents a low risk to life and can be discounted;

e. that there is a building on the far side of the boundary that has a similar elevation to the one in question, and that it is at the same distance from the common boundary; and

f. that the amount of radiation that passes through any part of the external wall that has fire resistance may be discounted.

14.3 Where a reduced separation distance is desired (or an increased amount of unprotected area) it may be advantageous to construct compartments of a smaller size.

Diagram 41 **Relevant boundary**

See paras 14.4 and 14.5

This diagram sets out the rules that apply in respect of a boundary for it to be considered as a relevant boundary

For a boundary to be relevant it should:
a coincide with, or
b be parallel to, or
c be at an angle of not more than 80° to the side of the building

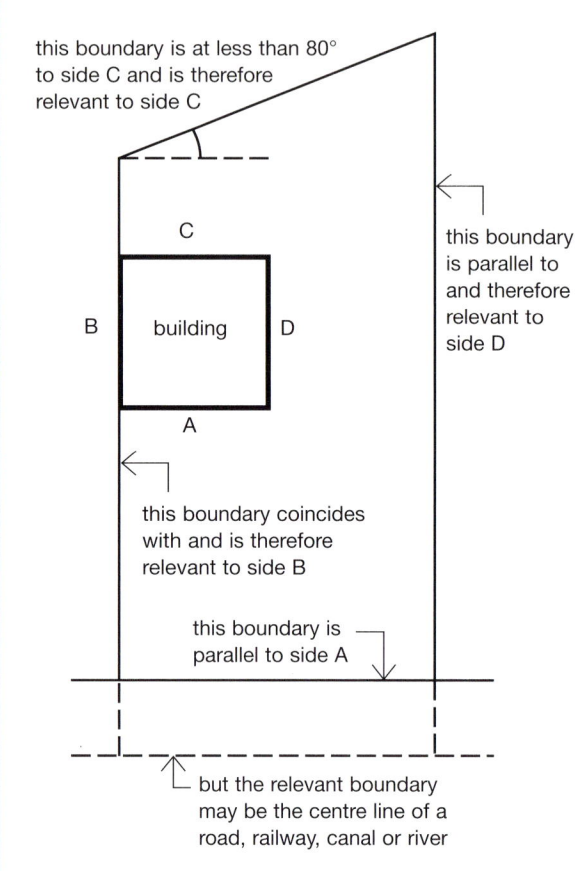

this boundary is at less than 80° to side C and is therefore relevant to side C

this boundary is parallel to and therefore relevant to side D

C

B   building   D

A

this boundary coincides with and is therefore relevant to side B

this boundary is parallel to side A

but the relevant boundary may be the centre line of a road, railway, canal or river

Boundaries

14.4 The use of the distance to a boundary, rather than to another building, in measuring the separation distance, makes it possible to calculate the allowable proportion of unprotected areas, regardless of whether there is a building on an adjoining site, and regardless of the site of that building, and the extent of any unprotected areas that it might have.

A wall is treated as facing a boundary if it makes an angle with it of 80° or less (see Diagram 41).

Usually only the distance to the actual boundary of the site needs to be considered. But in some circumstances, when the site boundary adjoins a space where further development is unlikely, such as a road, then part of the adjoining space may be included as falling within the relevant boundary for the purposes of this Section. The meaning of the term boundary is explained in Diagram 41.

Relevant boundaries

14.5 The boundary which a wall faces, whether it is the actual boundary of the site or a notional boundary, is called the relevant boundary (see Diagrams 41 and 42).

Notional boundaries

14.6 Generally separation distance between buildings on the same site is discounted. In some circumstances the distances to other buildings on the same site need to be considered. This is done by assuming that there is a boundary between those buildings. This assumed boundary is called a notional boundary. A notional boundary is assumed to exist where either or both of the buildings concerned are in the Residential or Assembly and Recreation Purpose Groups. The appropriate rules are given in Diagram 42.

Diagram 42 **Notional boundary**

See para 14.6

This diagram sets out the rules that apply where there is a building of the Residential or Assembly and Recreation Purpose Groups on the same site as another building, so that a notional boundary needs to be assumed between the buildings.

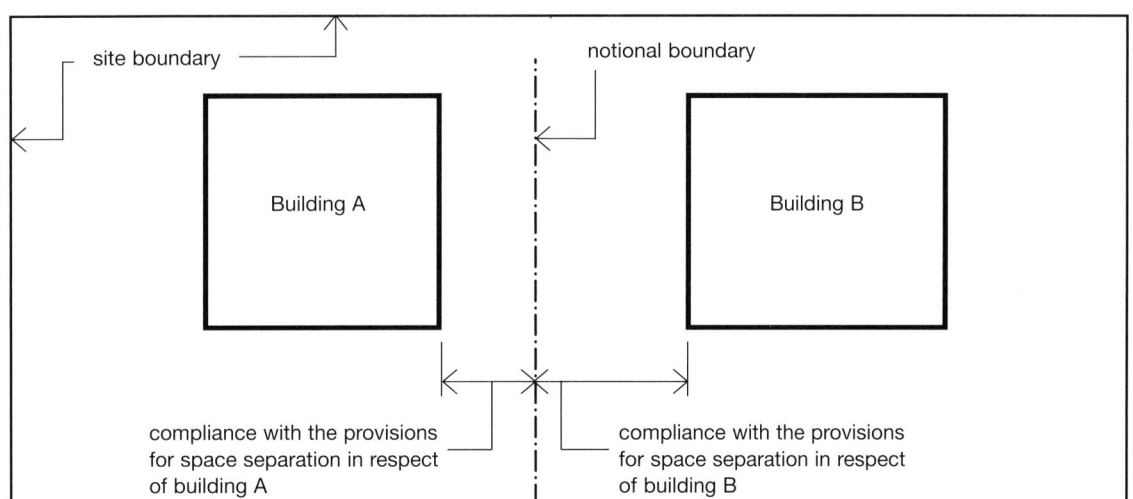

The notional boundary should be set in the area between the two buildings using the following rules:

1 It is only necessary to assume a notional boundary when the buildings are on the same site and either of the buildings, new or existing, is of Residential or Assembly and Recreation use.
2 The notional boundary is assumed to exist in the space between the buildings and is positioned so that one of the buildings would comply with the provisions for space separation having regard to the amount of its unprotected area. In practice, if one of the buildings is existing, the position of the boundary will be set by the space separation factors for that building.
3 The siting of the new building, or the second building if both are new, can then be checked to see that it also complies – using the notional boundary as the relevant boundary for the second building.

Unprotected areas

Unprotected areas and fire resistance

14.7 Any part of an external wall which has less fire resistance than the appropriate amount given in Appendix A, Table A2, is considered to be an unprotected area.

External walls of protected shafts forming stairways

14.8 Any part of an external wall of a stairway in a protected shaft is excluded from the assessment of unprotected area.

Note: There are provisions in the guidance to B1 (Diagram 21) and B5 (paragraph 18.11 which refers to Section 2 of BS 5588: Part 5: 1991 *Fire precautions in the design, construction and use of buildings, Code of practice for firefighting stairs and lifts*) about the relationship of external walls for protected stairways to the unprotected areas of other parts of the building.

Status of combustible surface materials as unprotected area

14.9 If an external wall has the appropriate fire resistance, but has combustible material more than 1mm thick as its external surface, then that wall is counted as an unprotected area amounting to half the actual area of the combustible material, see Diagram 43. (For the purposes of this provision, a material with a Class 0 rating (National class) or Class B-s3, d2 rating (European class) (see Appendix A, paragraphs 7 and 13) need not be counted as unprotected area).

Note: When a classification includes "s3, d2", this means that there is no limit set for smoke production and/or flaming droplets/particles.

Small unprotected areas

14.10 Small unprotected areas in an otherwise protected area of wall are considered to pose a negligible risk of fire spread, and may be desregarded. Diagram 44 shows the constraints that apply to the placing of such areas in relation to each other and to lines of compartmentation inside the building. These constraints vary according to the size of each unprotected area.

Canopies

14.11 Some canopy structures would be exempt from the application of the Building Regulations by falling within Class VI or Class VII of Schedule 2 to the Regulations (Exempt buildings and works). Many others may not meet the exemption criteria and in such cases the provisions in this Section about limits of unprotected areas could be onerous.

In the case of a canopy attached to the side of a building, provided that the edges of the canopy are at least 2m from the relevant boundary, separation distance may be determined from the wall rather than the edge of the canopy (see Diagram 45).

In the case of a free-standing canopy structure above a limited risk or controlled hazard (for example over petrol pumps), in view of the high degree of ventilation and heat dissipation achieved by the open sided construction, and provided the canopy is 1000mm or more from the relevant boundary, the provisions for space separation could reasonably be disregarded.

Large uncompartmented buildings

14.12 Parts of the external wall of an uncompartmented building which are more than 30m above mean ground level, may be disregarded in the assessment of unprotected area.

Diagram 43 **Status of combustible surface material as unprotected area**

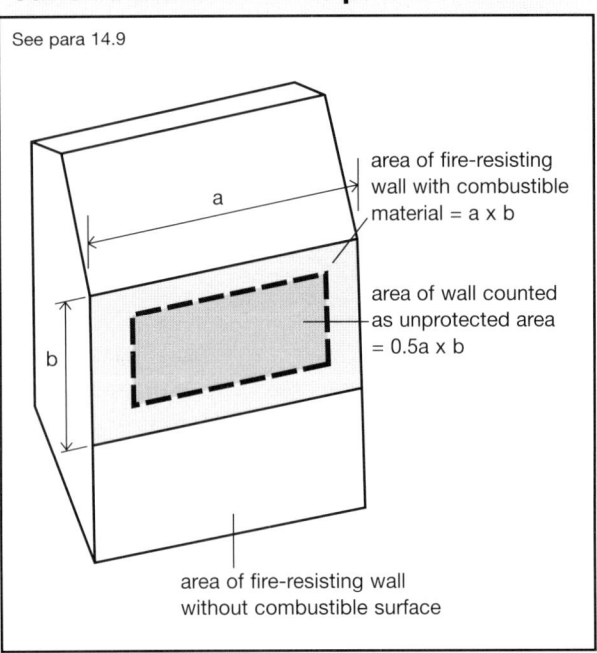

See para 14.9

a

b

area of fire-resisting wall with combustible material = a x b

area of wall counted as unprotected area = 0.5a x b

area of fire-resisting wall without combustible surface

Diagram 44 Unprotected areas which may be disregarded in assessing the separation distance from the boundary

See para 14.10

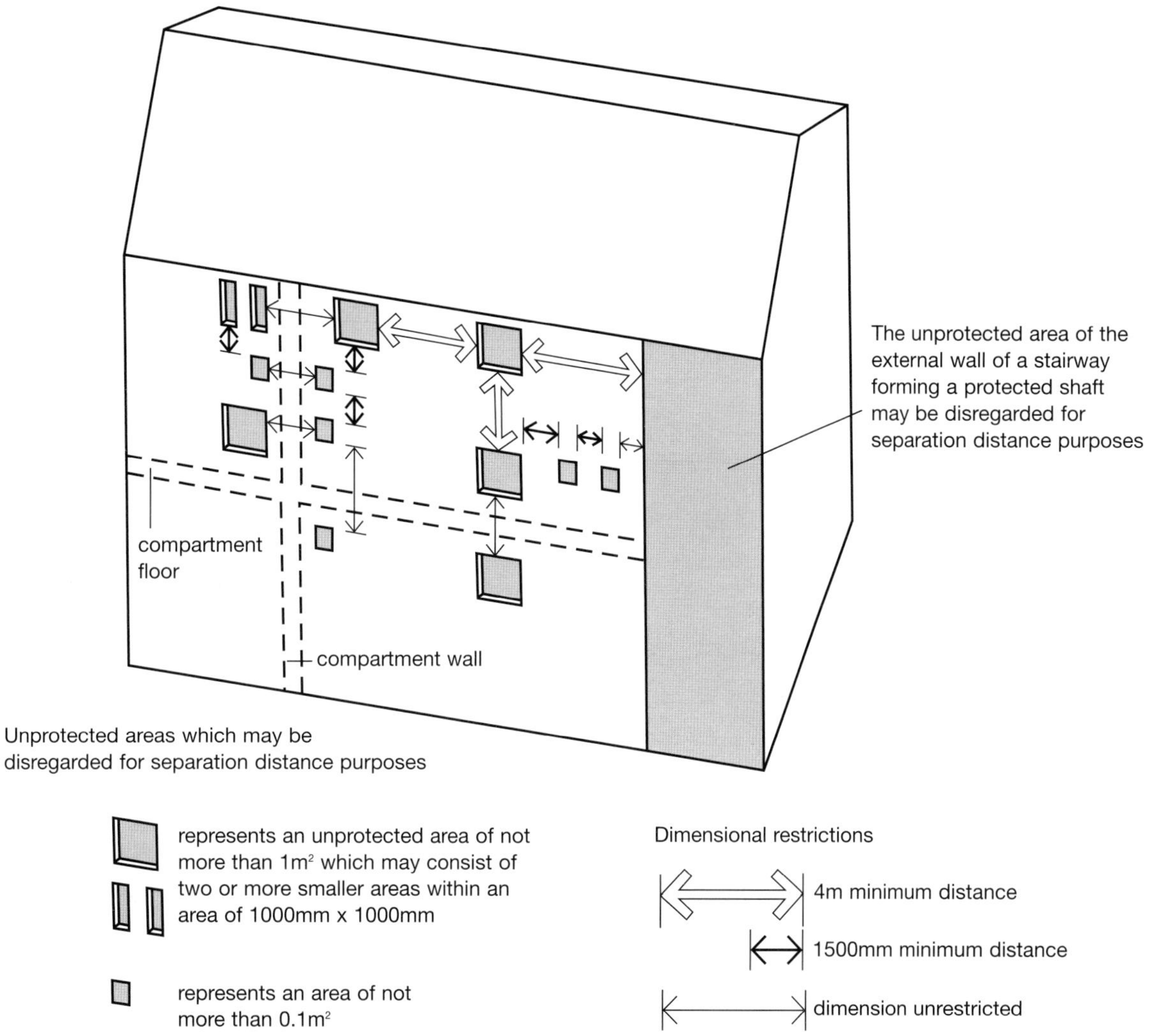

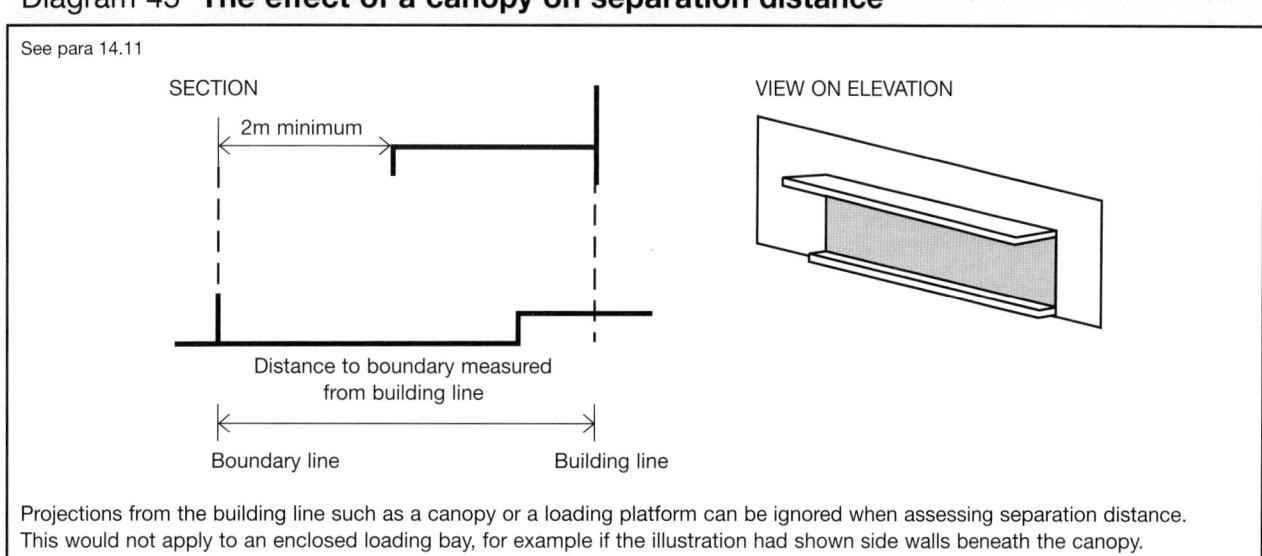

The unprotected area of the external wall of a stairway forming a protected shaft may be disregarded for separation distance purposes

compartment floor

compartment wall

Unprotected areas which may be disregarded for separation distance purposes

represents an unprotected area of not more than 1m² which may consist of two or more smaller areas within an area of 1000mm x 1000mm

represents an area of not more than 0.1m²

Dimensional restrictions

4m minimum distance

1500mm minimum distance

dimension unrestricted

Diagram 45 The effect of a canopy on separation distance

See para 14.11

SECTION

2m minimum

Distance to boundary measured from building line

Boundary line

Building line

VIEW ON ELEVATION

Projections from the building line such as a canopy or a loading platform can be ignored when assessing separation distance. This would not apply to an enclosed loading bay, for example if the illustration had shown side walls beneath the canopy.

External walls within 1000mm of the relevant boundary

14.13 A wall situated within 1000mm from any point on the relevant boundary, and including a wall coincident with the boundary, will meet the provisions for space separation if:

a. the only unprotected areas are those shown in Diagram 44 or referred to in paragraph 14.12; and

b. the rest of the wall is fire-resisting from both sides.

External walls 1000mm or more from the relevant boundary

14.14 A wall situated at least 1000mm from any point on the relevant boundary will meet the provisions for space separation if:

a. the extent of unprotected area does not exceed that given by one of the methods referred to in paragraph 14.15; and

b. the rest of the wall (if any) is fire-resisting.

Methods for calculating acceptable unprotected area

14.15 Two simple methods are given in this Approved Document for calculating the acceptable amount of unprotected area in an external wall that is at least 1000mm from any point on the relevant boundary. (For walls within 1000mm of the boundary see 14.13 above.)

Method 1 may be used for small residential buildings which do not belong to Purpose Group 2a (Institutional type premises), and is set out in paragraph 14.19.

Method 2 may be used for most buildings or compartments for which Method 1 is not appropriate, and is set out in paragraph 14.20.

There are other more precise methods, described in a BRE Report *External fire spread: Building separation and boundary distances* (BR 187, BRE 1991), which may be used instead of methods 1 and 2. The "Enclosing Rectangle" and "Aggregate Notional Area" methods are included in the BRE Report.

Basis for calculating acceptable unprotected area

14.16 The basis of Methods 1 and 2 is set out in Fire Research Technical Paper No 5, 1963. This has been reprinted as part of the BRE Report referred to in paragraph 14.15. The aim is to ensure that the building is separated from the boundary by at least half the distance at which the total thermal radiation intensity received from all unprotected areas in the wall would be 12.6 kw/m^2 (in still air), assuming the radiation intensity at each unprotected area is:

a. 84 kw/m^2, if the building is in the Residential, Office or Assembly and Recreation purpose groups, or is an open-sided multi-storey car park in Purpose Group 7(b); and

b. 168 kw/m^2, if the building is in the Shop and Commercial, Industrial, Storage or Other non-residential purpose groups.

Sprinkler systems

14.17 If a building is fitted throughout with a sprinkler system, it is reasonable to assume that the intensity and extent of a fire will be reduced. The sprinkler system should meet the relevant recommendations of BS 5306: Part 2 *Fire extinguishing installations and equipment on premises, Specification for sprinkler systems*, ie the relevant occupancy rating together with the additional requirements for life safety. In these circumstances the boundary distance may be half that for an otherwise similar, but unsprinklered, building, subject to there being a minimum distance of 1m. Alternatively, the amount of unprotected area may be doubled if the boundary distance is maintained.

Note: The presence of sprinklers may be taken into account in a similar way when using the BRE Report referred to in paragraph 14.15.

Atrium buildings

14.18 If a building contains one or more atria, the recommendations of clause 28.2 in BS 5588: Part 7: 1997 *Fire precautions in the design, construction and use of buildings, Code of practice for the incorporation of atria in buildings* should be followed.

Method 1 – Small Residential

14.19 This method applies only to a building intended to be used as a dwelling house, or for flats or other residential purposes (not Institutional), which is 1000mm or more from any point on the relevant boundary.

The following rules for determining the maximum unprotected area should be read with Diagram 46.

a. The building should not exceed 3 storeys in height (basements not counted) or be more than 24m in length:

b. Each side of the building will meet the provisions for space separation if:

    i.   the distance of the side of the building from the relevant boundary, and

    ii. the extent of the unprotected area,

    are within the limits given in Diagram 46.

    Note: In calculating the maximum unprotected area, any areas falling within the limits shown in Diagram 44, and referred to in paragraph 14.10, can be disregarded.

c. Any parts of the side of the building in excess of the maximum unprotected area should be fire-resisting.

Diagram 46 **Permitted unprotected areas in small residential buildings**

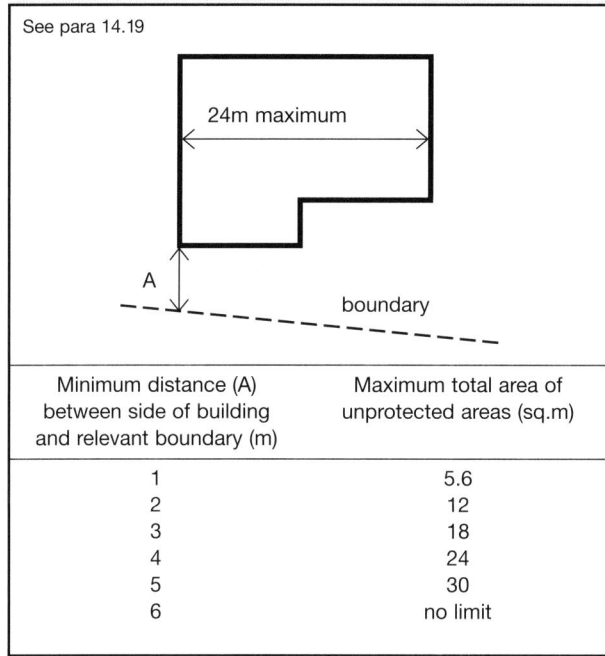

See para 14.19

24m maximum

A

boundary

Minimum distance (A) between side of building and relevant boundary (m)	Maximum total area of unprotected areas (sq.m)
1	5.6
2	12
3	18
4	24
5	30
6	no limit

Method 2 – Other buildings or compartments

14.20 This method applies to a building or compartment intended for any use and which is not less than 1000mm from any point on the relevant boundary. The following rules for determining the maximum unprotected area should be read with Table 16.

a. Except for an open-sided car park in Purpose Group 7(b) (see paragraph 12.4), the building or compartment should not exceed 10m in height.

    Note: For any building or compartment more than 10m in height, the methods set out in the BRE Report *External fire spread: Building separation and boundary distances* can be applied.

b. Each side of the building will meet the provisions for space separation if either:

    i.   the distance of the side of the building from the relevant boundary, and

    ii. the extent of unprotected area,

    are within the appropriate limits given in Table 16;

    Note: In calculating the maximum unprotected area, any areas shown in Diagram 44, and referred to in paragraph 14.10, can be disregarded.

c. any parts of the side of the building in excess of the maximum unprotected area should be fire-resisting.

Table 16 **Permitted unprotected areas in small buildings or compartments**

Minimum distance between side of building and relevant boundary (m)		Maximum total percentage of unprotected area %
Purpose groups		
Residential, Office, Assembly and Recreation	Shop & Commercial, Industrial, Storage & other Non-residential	
(1)	(2)	(3)
n.a	1	4
1	2	8
2.5	5	20
5	10	40
7.5	15	60
10	20	80
12.5	25	100

Notes:

n.a = not applicable

a. Intermediate values may be obtained by interpolation.

b. For buildings which are fitted throughout with an automatic sprinkler system, see para 14.17.

c. In the case of open-sided car parks in Purpose Group 7(b), the distances set out in column (1) may be used instead of those in column (2).

Section 15

ROOF COVERINGS

Introduction

15.1 The provisions in this section limit the use, near a boundary, of roof coverings which will not give adequate protection against the spread of fire over them. The term roof covering is used to describe constructions which may consist of one or more layers of material, but does not refer to the roof structure as a whole. The provisions in this Section are principally concerned with the performance of roofs when exposed to fire from the outside.

Note: Currently, no guidance is possible on the performance requirements in terms of the resistance of roof coverings to external fire exposure as determined by the methods specified in DD ENV 1187:2002, since there is no accompanying classification procedure and no supporting comparative data.

15.2 The circumstances when a roof is subject to the provisions in Section 14 for space separation are explained in paragraph 14.1.

Other controls on roofs

15.3 There are provisions concerning the fire properties of roofs in three other Sections of this Document. In the guidance to B1 (paragraph 6.3) there are provisions for roofs that are part of a means of escape. In the guidance to B2 (paragraph 7.13) there are provisions for the internal surfaces of rooflights as part of the internal lining of a room or circulation space. In the guidance to B3 there are provisions in Section 8 for roofs which are used as a floor, and in Section 9 for roofs that pass over the top of a compartment wall.

Classification of performance

15.4 The performance of roof coverings is designated by reference to the test methods specified in BS 476 *Fire tests on building materials and structures*, Part 3: 1958 *External fire exposure roof tests*, as described in Appendix A. The notional performance of some common roof coverings is given in Table A5 of Appendix A.

Rooflights are controlled on a similar basis, and plastics rooflights described in paragraph 15.6 and 15.7 may also be used.

Separation distances

15.5 The separation distance is the minimum distance from the roof (or part of the roof) to the relevant boundary, which may be a notional boundary.

Table 17 sets out separation distances according to the type of roof covering and the size and use of the building. There are no restrictions on the use of roof coverings designated AA, AB or AC.

Note: The boundary formed by the wall separating a pair of semi-detached houses may be disregarded for the purposes of this Section (but see Section 9, Diagram 28(b), which deals with roofs passing over the top of a compartment wall).

Diagram 47 Limitations on spacing and size of plastics rooflights having a Class 3 or TP(b) lower surface

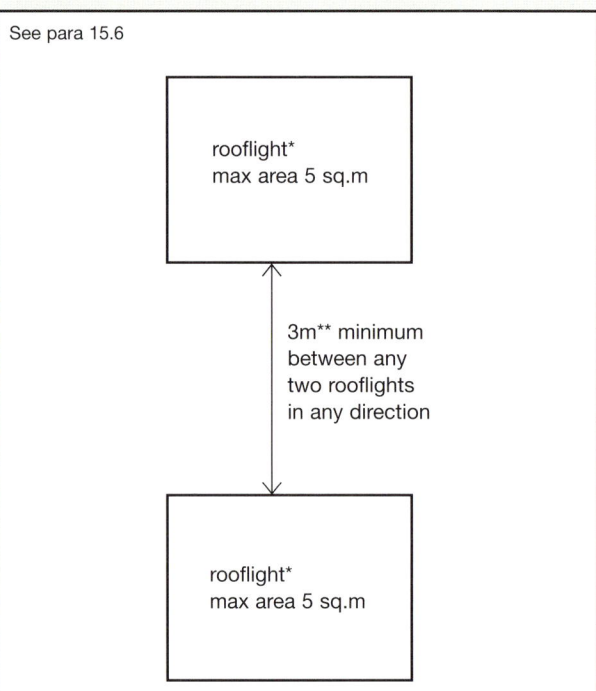

See para 15.6

rooflight*
max area 5 sq.m

3m** minimum between any two rooflights in any direction

rooflight*
max area 5 sq.m

*   or group of rooflights amounting to no more than 5 sq.m

**   class 3 rooflights to rooms in industrial and other non-residential purpose groups may be spaced 1800mm apart provided the rooflights are evenly distributed and do not exceed 20% of the area of the room

Notes:
1   there are restrictions on the use of plastic rooflights in the guidance to B2.
2   surrounding roof covering to be a material of limited combustibility for at least 3m distance.
3   where Diagram 28a or b applies, rooflights should be at least 1500mm from the compartment wall.

Plastics rooflights

15.6 Table 18 sets out the limitations on the use of plastics rooflights which have at least a Class 3 lower surface, and Table 19 sets out the limitations on the use of thermoplastic materials with a TP(a) rigid or TP(b) classification (see also Diagram 47). The method of classifying thermoplastic materials is given in Appendix A.

15.7 When used in rooflights, a rigid thermoplastic sheet product made from polycarbonate or from unplasticised PVC, which achieves a Class 1 rating for surface spread of flame when tested to BS 476 *Fire tests on building materials and structures*, Part 7: 1971 (or 1987 or 1997) *Surface spread of flame tests for materials*, can be regarded as having an AA designation.

Unwired glass in rooflights

15.8 When used in rooflights, unwired glass at least 4mm thick has an AA designation.

Thatch and wood shingles

15.9 Thatch and wood shingles should be regarded as having an AD/BD/CD designation in Table 17 if performance under BS 476: Part 3: 1958 cannot be established.

Note: Consideration can be given to thatched roofs being closer to the boundary than shown in Table 17 if, for example, the following precautions (based on *Thatched buildings. New properties and extensions* [the "Dorset Model"]) are incorporated in the design:

a. the rafters are overdrawn with construction having not less than 30 min fire resistance;

b. the guidance given in Approved Document J *Combustion appliances and fuel storage* is followed; and

c. the smoke alarm installation (see Section 1) is included in the roof space.

Table 17 **Limitations on roof coverings***

Designation† of covering of roof or part of roof	Minimum distance from any point on relevant boundary			
	Less than 6m	At least 6m	At least 12m	At least 20m
AA, AB or AC	●	●	●	●
BA, BB or BC	○	●	●	●
CA, CB or CC	○	● (1)(2)	● (1)	●
AD, BD or CD (1)	○	● (2)	●	●
DA, DB, DC or DD (1)	○	○	○	● (2)

Notes:

* See paragraph 15.8 for limitations on glass; paragraph 15.9 for limitations on thatch and wood shingles; and paragraphs 15.6 and 15.7 and Tables 18 and 19 for limitations on plastics rooflights.

† The designation of external roof surfaces is explained in Appendix A. (See Table A5, for notional designations of roof coverings.)

   Separation distances do not apply to the boundary between roofs of a pair of semi-detached houses (see 15.5) and to enclosed/ covered walkways. However, see Diagram 28 if the roof passes over the top of a compartment wall.

   Polycarbonate and PVC rooflights which achieve a Class 1 rating by test, see paragraph 15.7, may be regarded as having an AA designation.

● Acceptable.
○ Not acceptable.

1. Not acceptable on any of the following buildings:
   a   Houses in terraces of three or more houses;
   b. Industrial, Storage or Other non-residential purpose group buildings of any size;
   c. Any other buildings with a cubic capacity of more than 1500m³.

2. Acceptable on buildings not listed in Note 1, if part of the roof is no more than 3m² in area and is at least 1500mm from any similar part, with the roof between the parts covered with a material of limited combustibility.

Table 18 Class 3 plastics rooflights: limitations on use and boundary distance

Minimum classification on lower surface (1)	Space which rooflight can serve	Minimum distance from any point on relevant boundary to rooflight with an external designation† of:	
		AD BD CA CB CC CD	DA DB DC DD
Class 3	a. balcony, verandah, carport, covered way or loading bay, which has at least one longer side wholly or permanently open.	6m	20m
	b. detached swimming pool.		
	c. conservatory, garage or outbuilding, with a maximum floor area of 40m².		
	d. circulation space (2) (except a protected stairway).	6m (3)	20m (3)
	e. room (2).		

Notes:

† The designation of external roof surfaces is explained in Appendix A.

None of the above designations are suitable for protected stairways – see paragraph 7.13.

Polycarbonate and PVC rooflights which achieve a Class 1 rating by test, see paragraph 15.7, may be regarded as having an AA designation.

Where Diagram 28a or b applies, rooflights should be at least 1.5m from the compartment wall.

Products may have upper & lower surfaces with different properties if they have double skins or are laminates of different materials. In which case the more onerous distance applies.

1. See also the guidance to B2.
2. Single skin rooflight only, in the case of non-thermoplastic material.
3. The rooflight should also meet the provisions of Diagram 47.

Table 19 TP(a) and TP(b) plastics rooflights: limitations on use and boundary distance

Minimum classification on lower surface (1)	Space which rooflight can serve	Minimum distance from any point on relevant boundary to rooflight with an external surface classification (1) of:	
		TP(a)	TP(b)
1. TP(a) rigid	any space except a protected stairway	6m (2)	not applicable
2. TP(b)	a. balcony, verandah, carport, covered way or loading bay, which has at least one longer side wholly or permanently open.	not applicable	6m
	b. detached swimming pool.		
	c. conservatory, garage or outbuilding, with a maximum floor area of 40m².		
	d. circulation space (3) (except a protected stairway).	not applicable	6m (4)
	e. room (3).		

Notes:

None of the above designations are suitable for protected stairways – see paragraph 7.13.

Polycarbonate and PVC rooflights which achieve a Class 1 rating by test, see paragraph 15.7, may be regarded as having an AA designation.

Where Diagram 28a or b applies, rooflights should be at least 1.5m from the compartment wall.

Products may have upper & lower surfaces with different properties if they have double skins or are laminates of different materials. In which case the more onerous distance applies.

1. See also the guidance to B2.
2. No limit in the case of any space described in 2a, b & c.
3. Single skin rooflight only, in the case of non-thermoplastic material.
4. The rooflight should also meet the provisions of Diagram 47.

The Requirement

This Approved Document deals with the following Requirement from Part B of Schedule 1 to the Building Regulations 2000.

Requirement	Limits on application
Access and facilities for the fire service **B5.**-(1) The building shall be designed and constructed so as to provide reasonable facilities to assist fire fighters in the protection of life. (2) Reasonable provision shall be made within the site of the building to enable fire appliances to gain access to the building.	

Guidance

Performance

In the Secretary of State's view the requirement of B5 will be met:

a. if there is sufficient means of external access to enable fire appliances to be brought near to the building for effective use;

b. if there is sufficient means of access into, and within, the building for fire-fighting personnel to effect rescue and fight fire;

c. if the building is provided with sufficient internal fire mains and other facilities to assist fire-fighters in their tasks; and

d. if the building is provided with adequate means for venting heat and smoke from a fire in a basement.

These access arrangements and facilities are only required in the interests of the health and safety of people in and around the building. The extent to which they are required will depend on the use and size of the building in so far as it affects the health and safety of those people.

Introduction

B5.i The guidance given here covers the selection and design of facilities for the purpose of protecting life by assisting the fire service.

To assist the fire service some or all of the following facilities may be necessary, depending mainly on the size of the building:

a. vehicle access for fire appliances;

b. access for fire-fighting personnel;

c. the provision of fire mains within the building;

d. venting for heat and smoke from basement areas.

Facilities appropriate to a specific building

B5.ii The main factor determining the facilities needed to assist the fire service is the size of the building. Generally speaking fire fighting is carried out within the building.

a. In deep basements and tall buildings (see paragraph 18.2) firefighters will invariably work inside. They need special access facilities (see Section 18), equipped with fire mains (see Section 16). Fire appliances will need access to entry points near the fire mains (see Section 17).

b. In other buildings, the combination of personnel access facilities offered by the normal means of escape, and the ability to work from ladders and appliances on the perimeter, will generally be adequate without special internal arrangements. Vehicle access may be needed to some or all of the perimeter, depending on the size of the building (see Section 17);

c. For dwellings and other small buildings, it is usually only necessary to ensure that the building is suffiently close to a point accessible to fire brigade vehicles (see paragraph 17.2);

d. In taller blocks of flats, fire brigade personnel access facilities are needed within the building, although the high degree of compartmentation means that some simplification is possible compared to other tall buildings (see paragraph 18.12);

e. Products of combustion from basement fires tend to escape via stairways, making access difficult for fire service personnel. The problem can be reduced by providing vents (see Section 19). Venting can improve visibility and reduce temperatures, making search, rescue and fire-fighting less difficult.

Insulating core panels

B5.iii Guidance on the fire behaviour of insulating core panels used for internal structures is given in Appendix F.

Section 16

FIRE MAINS

Introduction

16.1 Fire mains are installed in a building and equipped with valves etc so that the fire service may connect hoses for water to fight fires inside the building. Rising fire mains serve floors above ground, or upwards from the level at which the fire service gain access (called the fire service vehicle access level) if this is not ground level. (In a podium design for instance, the fire service vehicle access level may be above the ground level, see Diagram 51.) Falling mains serve levels below fire service vehicle access level.

Fire mains may be of the 'dry' type which are normally empty and are supplied through hose from a fire service pumping appliance. Alternately they may be of the 'wet' type where they are kept full of water and supplied from tanks and pumps in the building. There should be a facility to allow a wet system to be replenished from a pumping appliance in an emergency.

Provision of fire mains

16.2 Buildings provided with firefighting shafts should be provided with fire mains in those shafts. The criteria for the provision of firefighting shafts are given in Section 18.

16.3 Wet rising mains should be provided in buildings with a floor at more than 60m above fire service vehicle access level. In lower buildings where fire mains are provided, either wet or dry mains are suitable.

Number and location of fire mains

16.4 There should be one fire main in every firefighting shaft. (See Section 18 for guidance on the provision of firefighting shafts.)

16.5 The outlets from fire mains in firefighting shafts should be sited in each firefighting lobby giving access to the accommodation. (See Section 18, paragraphs 18.9 and 18.10.)

Design and construction of fire mains

16.6 Guidance on other aspects of the design and construction of fire mains, not included in the provisions of this Approved Document, should be obtained from sections 2 and 3 of BS 5306: Part 1: 1976 *Fire extinguishing installations and equipment on premises, Hydrant systems, hose reels and foam inlets.*

Section 17

VEHICLE ACCESS

Introduction

17.1 For the purposes of this Approved Document vehicle access to the exterior of a building is needed to enable high reach appliances, such as turntable ladders and hydraulic platforms, to be used, and to enable pumping appliances to supply water and equipment for firefighting and rescue activities.

Access requirements increase with building size and height.

Fire mains (see Section 16) enable fire-fighters within the building to connect their hoses to a water supply. In buildings fitted with fire mains, pumping appliances need access to the perimeter at points near the mains, where fire-fighters can enter the building and where in the case of dry mains, a hose connection will be made from the appliance to pump water into the main.

The vehicle access requirements described in Table 20 for buildings without fire mains, do not apply to blocks of flats & maisonettes because access is required to each individual dwelling (see 17.3), or to buildings with fire mains.

Vehicle access routes and hard-standings should meet the criteria decribed in paragraphs 17.8 to 17.11 where they are to be used by fire service vehicles.

Note: Requirements cannot be made under the Building Regulations for work to be done outside the site of the works shown on the deposited plans, building notice or initial notice. In this connection it may not always be reasonable to upgrade an existing route across a site to a small building such as a single dwelling-house. The options in such a case, from doing no work to upgrading certain features of the route eg a sharp bend, should be considered by the Building Control Body in consultation with the fire service.

Buildings not fitted with fire mains

17.2 There should be vehicle access for a pump appliance to small buildings (those of up to 2000m² with a top storey up to 11m above ground level) to either:

a.   15% of the perimeter; or

b.   within 45m of every point on the projected plan area (or 'footprint', see Diagram 48) of the building;

whichever is the less onerous.

Note: For single family dwelling houses, the 45m may be measured to a door to the dwelling.

17.3 There should be vehicle access for a pump appliance to blocks of flats/maisonettes to within 45m of every dwelling entrance door.

17.4 Vehicle access to buildings that do not have fire mains (other than buildings described in paragraphs 17.2 and 17.3) should be provided in accordance with Table 20.

17.5 Every elevation to which vehicle access is provided in accordance with paragraphs 17.2 or 17.3 or Table 20 should have a suitable door, not less than 750mm wide, giving access to the interior of the building.

Table 20   Fire service vehicle access to buildings (excluding blocks of flats) not fitted with fire mains

Total floor area (1) of building m²	Height of floor of top storey above ground (2)	Provide vehicle access (3)(4) to:	Type of appliance
up to 2000	up to 11	see paragraph 17.2	pump
	over 11	15% of perimeter (5)	high reach
2000-8000	up to 11	15% of perimeter (5)	pump
	over 11	50% of perimeter (5)	high reach
8000-16,000	up to 11	50% of perimeter (5)	pump
	over 11	50% of perimeter (5)	high reach
16000-24,000	up to 11	75% of perimeter (5)	pump
	over 11	75% of perimeter (5)	high reach
over 24,000	up to 11	100% of perimeter (5)	pump
	over 11	100% of perimeter (5)	high reach

Notes:

1.   The total floor area is the aggregate of all floors in the building (excluding basements).

2.   In the case of Purpose Group 7(a) (storage) buildings, height should be measured to mean roof level, see Methods of Measurement in Appendix C.

3.   An access door is required to each such elevation (see paragraph 17.5).

4.   See paragraph 17.9 for meaning of access.

5.   Perimeter is decribed in Diagram 48.

Diagram 48 **Example of building footprint and perimeter**

See para 17.2 and Table 20

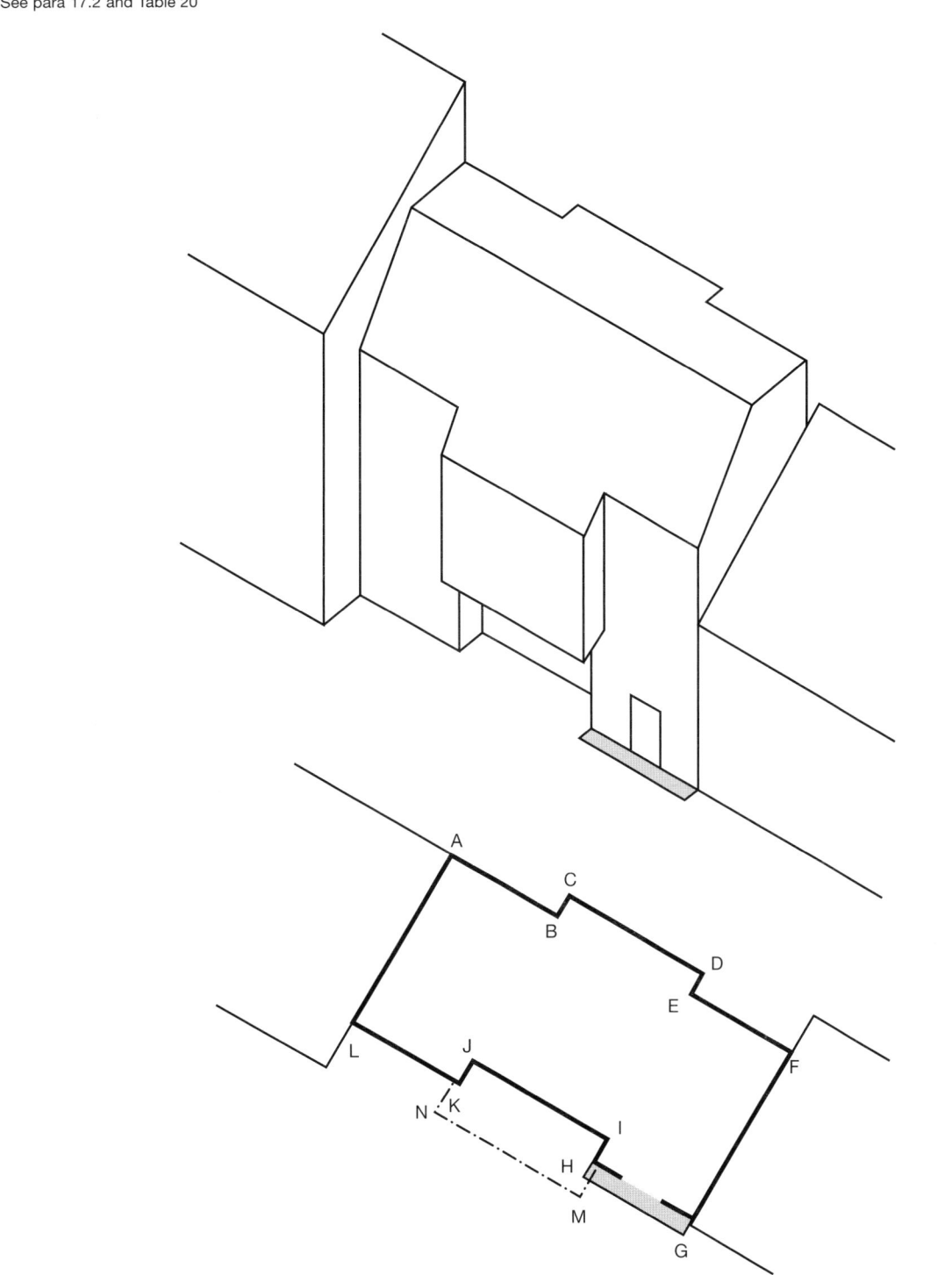

Plan of building AFGL where AL and FG are walls in common with other buildings.

The footprint of the building is the maximum aggregate plan perimeter found by the vertical projection of any overhanging storey onto a ground storey (ie ABCDEFGHMNKL).

The perimeter of the building for the purposes of Table 20 is the sum of the lengths of the two external walls, taking account of the footprint ie (A to B to C to D to E to F) + (G to H to M to N to K to L).

If the dimensions of the building were such that Table 20 requires vehicle access, the shaded area illustrates one possible example of 15% of the perimeter. **Note:** There should be a door into the building in this length (see paragraph 17.5).

If the building does not have walls in common with other buildings, the lengths AL and FG would be included in the perimeter.

Diagram 49   Relationship between building and hardstanding/access roads for high reach fire appliances

See para 17.10

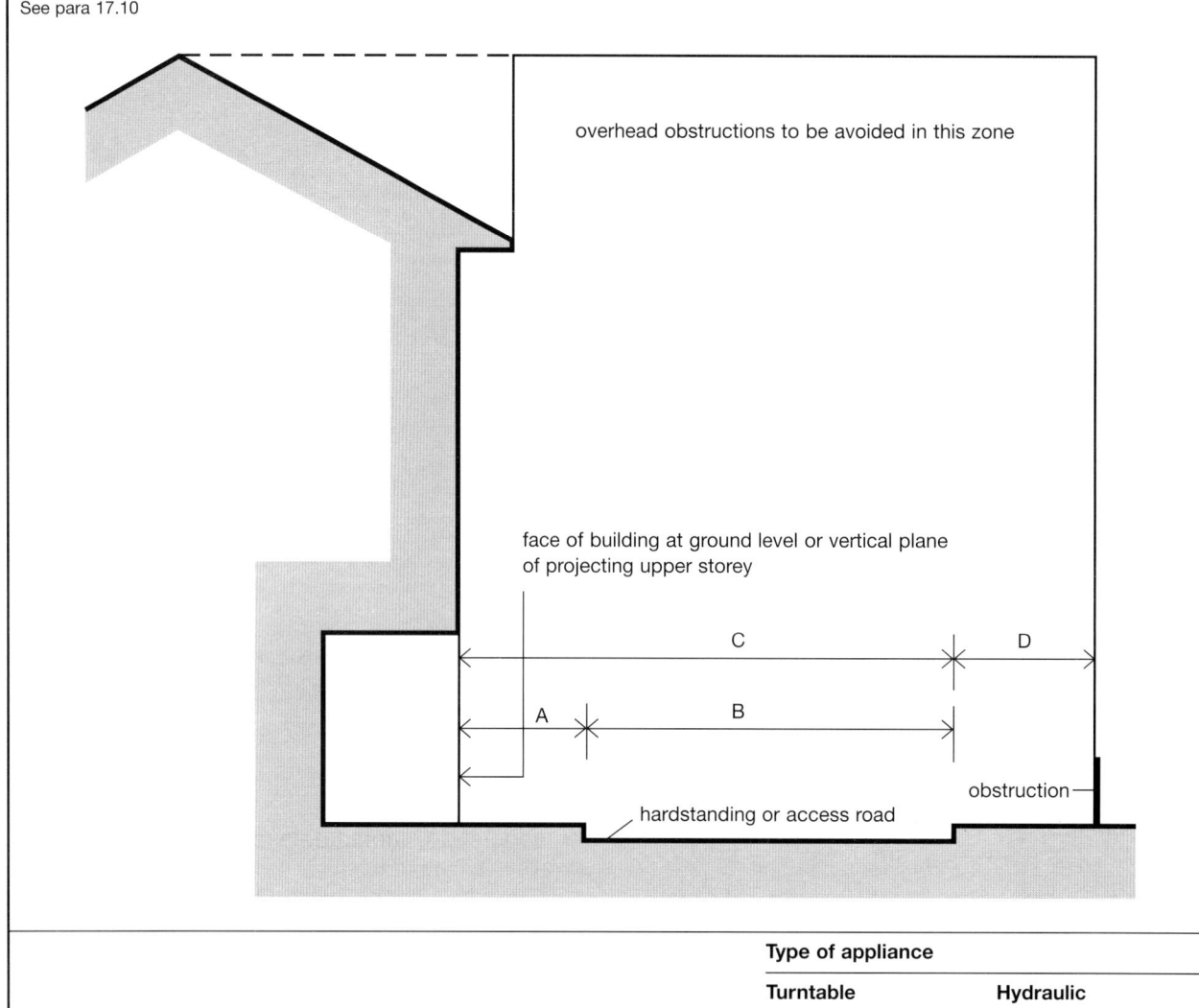

	Type of appliance	
	Turntable ladder Dimension (m)	**Hydraulic platform** Dimension (m)
A. Maximum distance of near edge of hardstanding from building	4.9	2.0
B. Minimum width of hardstanding	5.0	5.5
C. Minimum distance of further edge of hardstanding from building	10.0	7.5
D. Minimum width of unobstructed space (for swing of appliance platform)	NA	2.2

Notes:

1   Hardstanding for high reach appliances should be as level as possible and should not exceed a gradient of 1 in 12.

2   The hardstanding should be capable of withstanding a point load of 8.3 kg/cm^2 to accommodate jacks.

Table 21   Typical fire service vehicle access route specification

Appliance type	Minimum width of road between kerbs (m)	Minimum width of gateways (m)	Minimum turning circle between kerbs (m)	Minimum turning circle between walls (m)	Minimum clearance height (m)	Minimum carrying capacity (tonnes)
Pump	3.7	3.1	16.8	19.2	3.7	12.5
High reach	3.7	3.1	26.0	29.0	4.0	17.0

Notes:

1. Fire appliances are not standardised. Some fire services have appliances of greater weight or different size. In consultation with the Fire Authority, the Building Control Body may adopt other dimensions in such circumstances.

2. Because the weight of high reach appliances is distributed over a number of axles, it is considered that their infrequent use of a carriageway or route designed to 12.5 tonnes should not cause damage. It would therefore be reasonable to design the roadbase to 12.5 tonnes, although structures such as bridges should have the full 17 tonnes capacity.

Buildings fitted with fire mains

Note: Where fire mains are provided in buildings for which Sections 16 and 18 make no provision, vehicle access may be to paragraph 17.6 or 17.7 rather than Table 20.

17.6 In the case of a building fitted with dry fire mains there should be access for a pumping appliance to within 18m of each fire main inlet connection point. The inlet should be visible from the appliance.

17.7 In the case of a building fitted with wet mains the pumping appliance access should be to within 18m, and within sight of, a suitable entrance giving access to the main, and in sight of the inlet for the emergency replenishment of the suction tank for the main.

Design of access routes and hard standings

17.8 A vehicle access route may be a road or other route which, including any manhole covers and the like, meets the standards in Table 21 and the following paragraphs.

17.9 Where access is provided to an elevation in accordance with Table 20 for:

a. buildings up to 11m in height (excluding buildings covered by paragraphs 17.2(b) and 17.3), there should be access for a pump appliance adjacent to the building for the percentage of the total perimeter specified;

b. buildings over 11m in height, the access routes should meet the guidance in Diagram 49.

17.10 Where access is provided to an elevation for high reach appliances in accordance with Table 20, overhead obstructions such as cables and branches that would interfere with the setting of ladders etc, should be avoided in the zone shown in Diagram 49.

17.11 Turning facilities should be provided in any dead-end access route that is more than 20m long (see Diagram 50). This can be by a hammerhead or turning circle, designed on the basis of Table 21.

Diagram 50   Turning facilities

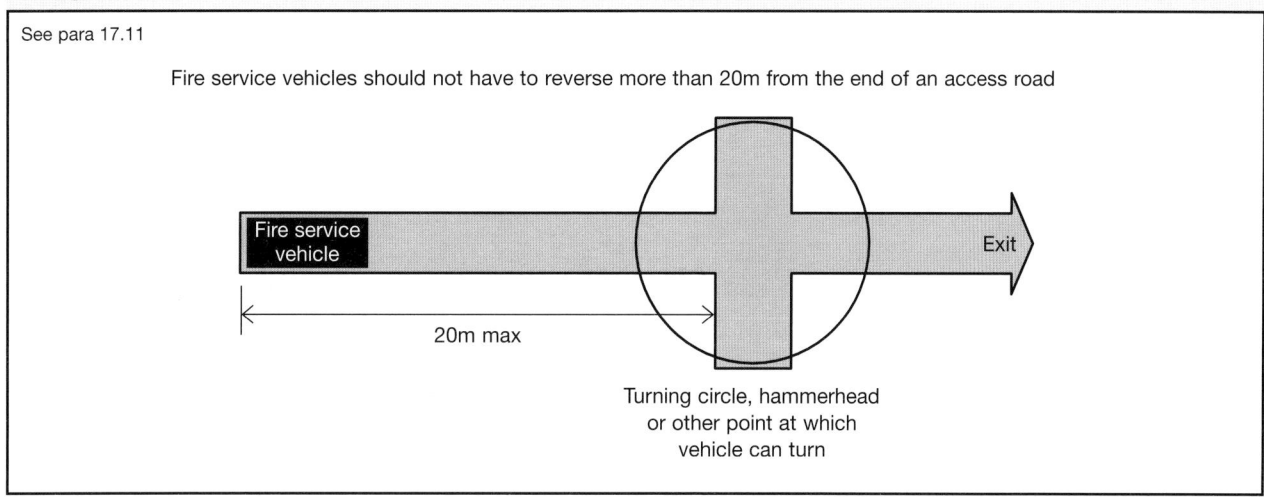

See para 17.11

Fire service vehicles should not have to reverse more than 20m from the end of an access road

Fire service vehicle

Exit

20m max

Turning circle, hammerhead or other point at which vehicle can turn

Section 18

ACCESS TO BUILDINGS FOR FIRE-FIGHTING PERSONNEL

Introduction

18.1 In low rise buildings without deep basements fire service personnel access requirements will be met by a combination of the normal means of escape, and the measures for vehicle access in Section 17, which facilitate ladder access to upper storeys. In other buildings the problems of reaching the fire, and working inside near the fire, necessitate the provision of additional facilities to avoid delay and to provide a sufficiently secure operating base to allow effective action to be taken.

These additional facilities include firefighting lifts, firefighting stairs and firefighting lobbies, which are combined in a protected shaft known as the firefighting shaft (Diagram 52).

Guidance on protected shafts in general is given in Section 9.

Note: Because of the high degree of compartmentation in blocks of flats/ maisonettes, the provisions in this Section may be modified (see paragraph 18.12).

Provision of firefighting shafts

18.2 Buildings with a floor at more than 18m above fire service vehicle access level, or with a basement at more than 10m below fire service vehicle access level, should be provided with firefighting shafts containing firefighting lifts (see Diagram 51).

18.3 Buildings in Purpose Groups 4, 6 and 7(a) with a storey of 900m² or more in area, where the floor is at a height of more than 7.5m above fire service vehicle access level, should be provided with firefighting shaft(s), which need not include firefighting lifts.

18.4 Buildings with two or more basement storeys, each exceeding 900m² in area, should be provided with firefighting shaft(s), which need not include firefighting lifts.

18.5 If a firefighting shaft is required to serve a basement it need not also serve the upper floors unless they also qualify because of the height or size of the building. Similarly a shaft serving upper storeys need not serve a basement which is not large or deep enough to qualify in its own right. However a firefighting stair and any firefighting lift should serve all intermediate storeys between the highest and lowest storeys that they serve.

Diagram 51   **Provision of firefighting shafts**

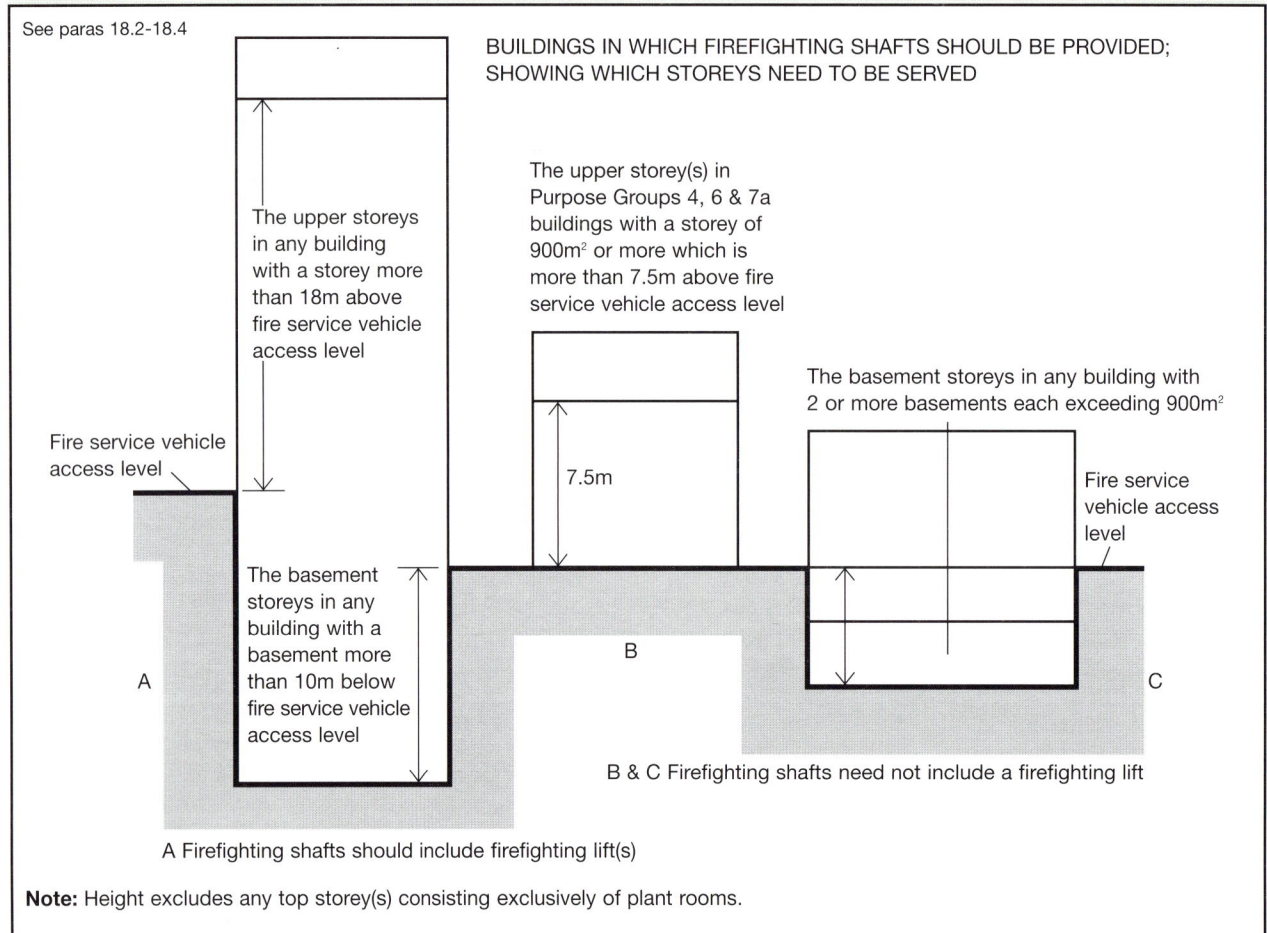

See paras 18.2-18.4

BUILDINGS IN WHICH FIREFIGHTING SHAFTS SHOULD BE PROVIDED; SHOWING WHICH STOREYS NEED TO BE SERVED

The upper storeys in any building with a storey more than 18m above fire service vehicle access level

The upper storey(s) in Purpose Groups 4, 6 & 7a buildings with a storey of 900m² or more which is more than 7.5m above fire service vehicle access level

The basement storeys in any building with 2 or more basements each exceeding 900m²

Fire service vehicle access level

Fire service vehicle access level

The basement storeys in any building with a basement more than 10m below fire service vehicle access level

7.5m

A

B

C

A Firefighting shafts should include firefighting lift(s)

B & C Firefighting shafts need not include a firefighting lift

Note: Height excludes any top storey(s) consisting exclusively of plant rooms.

Diagram 52 **Components of a firefighting shaft**

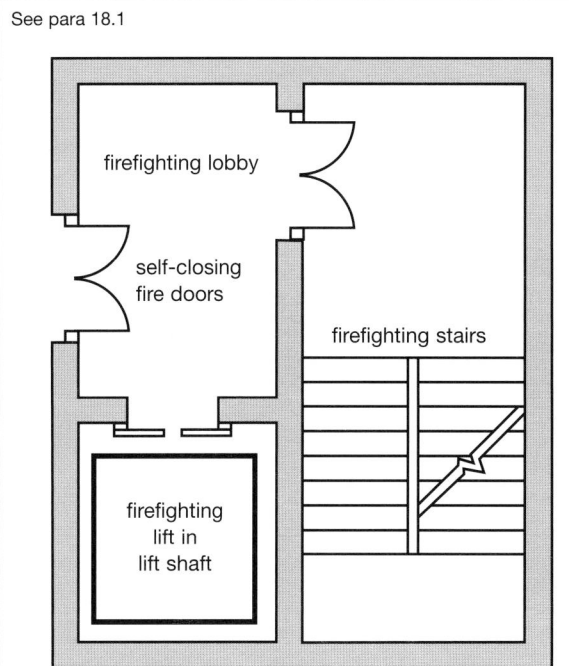

See para 18.1

firefighting lobby

self-closing fire doors

firefighting stairs

firefighting lift in lift shaft

Notes:

1. Outlets from a fire main should be located in the firefighting lobby.
2. A firefighting lift is required if the building has a floor more than 18m above, or more than 10m below, fire service vehicle access level.
3. This Diagram is only to illustrate the basic components and is not meant to represent the only acceptable layout. Ventilation measures have not been shown (refer to BS 5588: Part 5 *Code of practice for firefighting stairs and lifts*).

18.6 Shopping complexes should be provided with firefighting shafts in accordance with the recommendations of Section 3 of BS 5588: Part 10:
1991 *Fire precautions in the design, construction and use of buildings, Code of practice for shopping complexes*.

Number and location of firefighting shafts

18.7 The number of firefighting shafts should:

a. comply with Table 22, if the building is fitted throughout with an automatic sprinkler system meeting the relevant recommendations of BS 5306: Part 2 *Fire extinguishing installations and equipment on premises, Specification for sprinkler systems*; or

b. if the building is not fitted with sprinklers, be such that there is at least one for every 900m² (or part thereof) of floor area of the largest floor that is more than 18m above fire service vehicle access level (or above 7.5m covered by paragraph 18.3);

c. the same 900m² per firefighting shaft criterion should be applied to calculate the number of shafts needed where basements require them.

Table 22 **Minimum number of firefighting shafts in buildings fitted with sprinklers**

Largest qualifying floor area (m²)	Minimum number of firefighting shafts
less than 900	1
900-2000	2
over 2000	2 plus 1 for every additional 1500m² or part thereof

18.8 Firefighting shafts provided in accordance with paragraph 18.7 should be located such that every part of every storey, other than fire service access level, is no more than 60m from the fire main outlet, measured on a route suitable for laying hose. If the internal layout is unknown at the design stage, then every part of every such storey should be no more than 40m in a direct line from the fire main outlet.

Design and construction of firefighting shafts

18.9 Except in blocks of flats and maisonettes (see paragraph 18.12), every firefighting stair and firefighting lift should be approached from the accommodation, through a firefighting lobby.

18.10 All firefighting shafts should be equipped with fire mains having outlet connections and valves in every firefighting lobby.

18.11 Firefighting shafts should be designed, constructed and installed in accordance with the recommendations of BS 5588: Part 5: 1991 *Code of practice for firefighting stairs and lifts* in respect of the following:

a. Section 2: Planning and construction;

b. Section 3: Firefighting lift installation;

c. Section 4: Electrical services.

18.12 Where the design of means of escape in case of fire and compartmentation in blocks of flats/maisonettes has followed the guidance in Sections 3 and 9, the addition of a firefighting lobby between the firefighting stair(s) and the protected corridor or lobby provided for means of escape purposes is not necessary. Similarly, the firefighting lift can open directly into such protected corridor or lobby, but the firefighting lift landing doors should not be more than 7.5m from the door to the firefighting stair.

Rolling shutters in compartment walls

18.13 Rolling shutters should be capable of being opened and closed manually by the fire service.

Section 19

VENTING OF HEAT AND SMOKE FROM BASEMENTS

Introduction

19.1 The build-up of smoke and heat as a result of a fire can seriously inhibit the ability of the fire service to carry out rescue and firefighting operations in a basement. The problem can be reduced by providing facilities to make conditions tenable for firefighters.

19.2 Smoke outlets (also referred to as smoke vents) provide a route for heat and smoke to escape to the open air from the basement level(s). They can also be used by the fire service to let cooler air into the basement(s). (See Diagram 53.)

Provision of smoke outlets

19.3 Where practicable each basement space should have one or more smoke outlets, but it is not always possible to do this where, for example, the plan is deep and the amount of external wall is restricted by adjoining buildings. It is therefore acceptable to vent spaces on the perimeter and allow other spaces to be vented indirectly by fire-fighters opening connecting doors. However if a basement is compartmented, each compartment should have direct access to venting, without having to open doors etc into another compartment.

19.4 Smoke outlets, connected directly to the open air, should be provided from every basement storey, except for:

a. a basement in a single family dwellinghouse of Purpose Group 1(b) or 1(c); or

b. any basement storey that has:

    i. a floor area of not more than 200m^2, and

    ii. a floor not more than 3m below the adjacent ground level.

19.5 Strong rooms need not be provided with smoke outlets.

Diagram 53 Fire-resisting construction for smoke outlet shafts

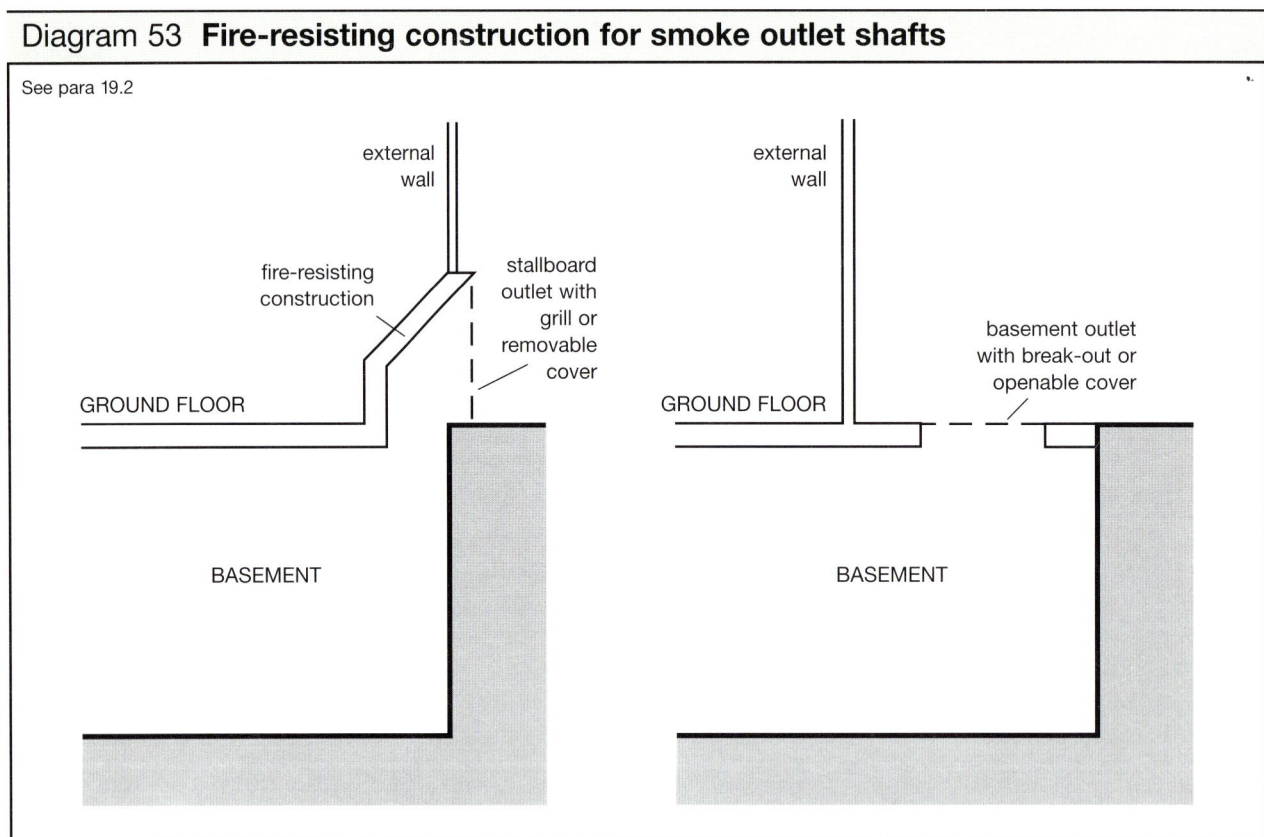

See para 19.2

external wall

fire-resisting construction

stallboard outlet with grill or removable cover

GROUND FLOOR

BASEMENT

external wall

basement outlet with break-out or openable cover

GROUND FLOOR

BASEMENT

19.6 Where basements have external doors or windows, the compartments containing the rooms with these doors or windows do not need smoke outlets. It is common for basements to be open to the air on one or more elevations. This may be the result of different ground levels on different sides of the building. It is also common in 18th and 19th century terraced housing where an area below street level is excavated at the front and/or rear of the terrace so that the lowest storey has ordinary windows, and sometimes an external door.

Natural smoke outlets

19.7 Smoke outlets should be sited at high level, either in the ceiling or in the wall of the space they serve. They should be evenly distributed around the perimeter to discharge in the open air outside the building.

19.8 The combined clear cross-sectional area of all smoke outlets should not be less than 1/40th of the floor area of the storey they serve.

19.9 Separate outlets should be provided from places of special fire hazard.

19.10 If the outlet terminates at a point that is not readily accessible, it should be kept unobstructed and should only be covered with a non-combustible grille or louvre.

19.11 If the outlet terminates in a readily accessible position, it may be covered by a panel, stallboard or pavement light which can be broken out or opened. The position of such covered outlets should be suitably indicated.

19.12 Outlets should not be placed where they would prevent the use of escape routes from the building.

Mechanical smoke extract

19.13 A system of mechanical extraction may be provided as an alternative to natural venting to remove smoke and heat from basements, provided that the basement storey(s) are fitted with a sprinkler system. The sprinkler system should be in accordance with the principles of BS 5306: Part 2 *Fire extinguishing installations and equipment on premises, Specification for sprinkler systems*. (It is not considered necessary in this particular case to install sprinklers on the storeys other than the basement(s) unless they are needed for other reasons.)

Note: Car parks are not normally expected to be fitted with sprinklers (see paragraph 12.2).

19.14 The air extraction system should give at least 10 air changes per hour and should be capable of handling gas temperatures of 300°C for not less than one hour. It should come into operation automatically on activation of the sprinkler system; alternatively activation may be by an automatic fire detection system which conforms to BS 5839: Part 1 *Fire detection and alarm systems for buildings, Code of practice for system design, installation and servicing* (at least L3 standard). For further information on equipment for removing hot smoke refer to BS 7346: Part 2 *Components for smoke and heat control systems, Specification for powered smoke and heat exhaust ventilators*.

Construction of outlet ducts or shafts

19.15 Outlet ducts or shafts, including any bulkheads over them (see Diagram 53), should be enclosed in non-combustible construction having not less fire resistance than the element through which they pass.

19.16 Where there are natural smoke outlet shafts from different compartments of the same basement storey, or from different basement storeys, they should be separated from each other by non-combustible construction having not less fire resistance than the storey(s) they serve.

Basement car parks

19.17 The provisions for ventilation of basement car parks in Section 12 may be taken as satisfying the requirements in respect of the need for smoke venting from any basement that is used as a car park.

Appendix A

PERFORMANCE OF MATERIALS AND STRUCTURES

Introduction

1 Much of the guidance in this document is given in terms of performance in relation to British or European Standards for products or methods of test or design or in terms of European Technical Approvals. In such cases the material, product or structure should:

a. be in accordance with a specification or design which has been shown by test to be capable of meeting that performance; or

Note: For this purpose, laboratories accredited by the United Kingdom Accreditation Service (UKAS) for conducting the relevant tests would be expected to have the necessary expertise.

b. have been assessed from test evidence against appropriate standards, or by using relevant design guides, as meeting that performance; or

Note: For this purpose, laboratories accredited by UKAS for conducting the relevant tests and suitably qualified fire safety engineers might be expected to have the necessary expertise.

For materials/products where European standards or approvals are not yet available and for a transition period after they become available, British standards may continue to be used. Any body notified to the UK Government by the Government of another member state of the European Union as capable of assessing such materials/products against the relevant British Standards, may also be expected to have the necessary expertise. Where European materials/products standards or approvals are available, any body notified to the European Commission as competent to assess such materials or products against the relevant European standards or technical approval can be considered to have the appropriate expertise.

c. where tables of notional performance are included in this document, conform with an appropriate specification given in these tables; or

d. in the case of fire-resisting elements:

i. conform with an appropriate specification given in Part II of the Building Research Establishments' Report *Guidelines for the construction of fire resisting structural elements* (BR 128, BRE 1988); or

ii. be designed in accordance with a relevant British Standard or Eurocode.

Note: Any test evidence used to substantiate the fire resistance rating of a construction should be carefully checked to ensure that it demonstrates compliance that is adequate and applicable to the intended use. Small differences in detail (such as fixing method, joints, dimensions, etc) may significantly affect the rating.

2 Building Regulations deal with fire safety in buildings as a whole. Thus they are aimed at limiting fire hazard.

The aim of standard fire tests is to measure or assess the response of a material, product, structure or system to one or more aspects of fire behaviour. Standard fire tests cannot normally measure fire hazard. They form only one of a number of factors that need to be taken into account. Other factors are set out in this publication.

Fire resistance

3 Factors having a bearing on fire resistance, that are considered in this document, are:

a. fire severity;

b. building height; and

c. building occupancy.

4 The standards of fire resistance given are based on assumptions about the severity of fires and the consequences should an element fail. Fire severity is estimated in very broad terms from the use of the building (its purpose group), on the assumption that the building contents (which constitute the fire load) are similar for buildings in the same use.

A number of factors affect the standard of fire resistance specified. These are:

a. the amount of combustible material per unit of floor area in various types of building (the fire load density);

b. the height of the top floor above ground, which affects the ease of escape and of fire fighting operations, and the consequences should large scale collapse occur;

c. occupancy type, which reflects the ease with which the building can be evacuated quickly;

d. whether there are basements, because the lack of an external wall through which to vent heat and smoke may increase heat build-up, and thus affect the duration of a fire, as well as complicating fire-fighting; and

e. whether the building is of single storey construction (where escape is direct and structural failure is unlikely to precede evacuation).

Because the use of buildings may change, a precise estimate of fire severity based on the fire load due to a particular use may be misleading. Therefore if a fire engineering approach of this kind is adopted the likelihood that the fire load may change in the future needs to be considered.

5 Performance in terms of the fire resistance to be met by elements of structure, doors and other forms of construction is determined by reference to either:

a. (National tests) BS 476 *Fire tests on building materials and structures*, Parts 20-24: 1987, ie Part 20 *Method for determination of the fire resistance of elements of construction (general principles),* Part 21 *Methods for determination of the fire resistance of loadbearing elements of construction*, Part 22 *Methods for determination of the fire resistance of non-loadbearing elements of construction*, Part 23 *Methods for determination of the contribution of components to the fire resistance of a structure*, and Part 24 *Method for determination of the fire resistance of ventilation ducts* (or to BS 476: Part 8: 1972 in respect of items tested or assessed prior to 1 January 1988); or

b. (European tests) Commission Decision 2000/367/EC of 3rd May 2000 implementing Council Directive 89/106/EEC as regards the classification of the resistance to fire performance of construction products, construction works and parts thereof.

Note: The designation of xxxx is used for the year reference for standards that are not yet published. The latest version of any standard may be used provided that it continues to address the relevant requirements of the Regulations.

All products are classified in accordance with BS EN 13501-2:xxxx, *Fire classification of construction products and building elements, Part 2- Classification using data from fire resistance tests (excluding products for use in ventilation systems).*

BS EN 13501- 3:xxxx, *Fire classification of construction products and building elements, Part 3-Classification using data from fire resistance tests on components of normal building service installations (other than smoke control systems).*

BS EN 13501-4:xxxx, *Fire classification of construction products and building elements, Part 4-Classification using data from fire resistance tests on smoke control systems.*

The relevant European test methods under BS EN 1364, 1365, 1366 and 1634 are listed in Appendix G.

Table A1 gives the specific requirements for each element in terms of one or more of the following performance criteria:

a. **resistance to collapse** (loadbearing capacity), which applies to loadbearing elements only, denoted R in the European classification of the resistance to fire performance;

b. **resistance to fire penetration** (integrity), denoted E in the European classification of the resistance to fire performance; and

c. **resistance to the transfer of excessive heat** (insulation), denoted I in the European classification of the resistance to fire performance.

Table A2 sets out the minimum periods of fire resistance for elements of structure.

Table A3 sets out criteria appropriate to the suspended ceilings that can be accepted as contributing to the fire resistance of a floor.

Table A4 sets out limitations on the use of uninsulated fire-resisting glazed elements. These limitations do not apply to the use of insulated fire-resisting glazed elements.

Information on tested elements is frequently given in literature available from manufacturers and trade associations.

Information on tests on fire-resisting elements is also given in such publications as:

Association for Specialist Fire Protection/Steel Construction Institute/Fire Test Study Group *Fire protection for structural steel in buildings*, second edition – revised, 1992. (Available from the ASFP, Association House, 99 West Street, Farnham, Surrey GU9 7EN and the Steel Construction Institute, Silwood Park, Ascot, Berks SL5 7QN).

Roofs

6 Performance in terms of the resistance of roofs to external fire exposure is determined by reference to the methods specified in BS 476: Part 3: 1958 *External fire exposure roof tests* under which constructions are designated by 2 letters in the range A to D, with an AA designation being the best. The first letter indicates the time to penetration, and the second letter a measure of the spread of flame.

Note: This is not the most recent version of the standard.

Currently, no guidance is possible on the performance in terms of the resistance of roofs to external fire exposure as determined by the methods specified in DD ENV 1187:2002, since there is no accompanying classification procedure and no comparative supporting data.

In some circumstances roofs, or parts of roofs, may need to be fire-resisting, for example if used as an escape route or if the roof performs the function of a floor. Such circumstances are covered in Sections 2, 6 and 8.

Table A5 gives notional designations of some generic roof coverings.

Reaction to fire

7 Performance in terms of reaction to fire to be met by construction products is determined by Commission Decision 200/147/EC of 8th February 2000 implementing Council Directive 89/106/EEC as regards the classification of the reaction to fire performance of construction products.

Note: The designation of xxxx is used for the year reference for standards that are not yet published. The latest version of any standard may be used provided that it continues to address the relevant requirements of the Regulations.

All products, excluding floorings, are classified as [†]A1, A2, B, C, D, E or F (with class A1 being the highest performance and F being the lowest) in accordance with BS EN 13501-1:2002, *Fire classification of construction products and building elements, Part 1-Classification using data from reaction to fire tests.*

The relevant European test methods are specified as follows,

BS EN ISO 1182:2002, *Reaction to fire tests for building products – Non-combustibility test.*

BS EN ISO 1716:2002 *Reaction to fire tests for building products – Determination of the gross calorific value.*

BS EN 13823:2002, *Reaction to fire tests for building products – Building products excluding floorings exposed to the thermal attack by a single burning item.*

BS EN ISO 11925-2:2002, *Reaction to fire tests for building Products, Part 2-Ignitability when subjected to direct impingement of a flame.*

BS EN 13238:2001, *Reaction to fire tests for building products-conditioning procedures and general rules for selection of substrates.*

Non-combustible materials

8 Non-combustible materials are defined in Table A6 either as listed products, or in terms of performance:

a. (National classes) when tested to BS 476: Part 4: 1970 *Non-combustibility test for materials* or Part 11: 1982 *Method for assessing the heat emission from building products;* or

b. (European classes) when classified as class A1 in accordance with BS EN 13501-1:2002, *Fire classification of construction products and building elements, Part 1-Classification using data from reaction to fire tests* when tested to BS EN ISO 1182:2002, *Reaction to fire tests for building products – Non-combustibility test* **and** BS EN ISO 1716:2002 *Reaction to fire tests for building products – Determination of the gross calorific value.*

Table A6 identifies non-combustible products and materials, and lists circumstances where their use is necessary.

Materials of limited combustibility

9 Materials of limited combustibility are defined in Table A7:

a. (National classes) by reference to the method specified in BS 476: Part 11: 1982; or

b. (European classes) in terms of performance when classified as class A2-s3, d2 in accordance with BS EN 13501-1:2002, *Fire classification of construction products and building elements, Part 1-Classification using data from reaction to fire tests* when tested to BS EN ISO 1182:2002, *Reaction to fire tests for building products – Non-combustibility test* **or** BS EN ISO 1716:2002 *Reaction to fire tests for building products – Determination of the gross calorific value* **and** BS EN 13823:2002, *Reaction to fire tests for building products – Building products excluding floorings exposed to the thermal attack by a single burning item.*

Table A7 also includes composite products (such as plasterboard) which are considered acceptable, and where these are exposed as linings they should also meet any appropriate flame spread rating.

Internal linings

10 Flame spread over wall or ceiling surfaces is controlled by providing for the lining materials or products to meet given performance levels in tests appropriate to the materials or products involved.

11 Under the National classifications, lining systems which can be effectively tested for 'surface spread of flame' are rated for performance by reference to the method specified in BS 476: Part 7: 1971 *Surface spread of flame tests for materials*, or 1987 *Method for classification of the surface spread of flame of products*, or 1997 *Method of test to determine the classification of the surface spread of flame of products* under which materials or products are classified 1, 2, 3 or 4 with Class 1 being the highest.

Under the European classifications, lining systems are classified in accordance with BS EN 13501-1:2002, *Fire classification of construction products and building elements, Part 1-Classification using data from reaction to fire tests.* Materials or products are classified as A1, A2, B, C, D, E or F, with A1 being the highest. When a classification includes "s3, d2", it means that there is no limit set for smoke production and/or flaming droplets/particles.

12 To restrict the use of materials which ignite easily, which have a high rate of heat release and/or which reduce the time to flash over, maximum acceptable 'fire propagation' indices are specified, where the National test methods are being followed. These are determined by reference to the method specified in BS 476: Part 6: 1981 or 1989 *Method of test for fire propagation of products.* Index of performance (I) relates to the overall test performance, whereas sub-index (i_i) is derived from the first three minutes of test.

13 The highest National product performance classification for lining materials is Class 0. This is achieved if a material or the surface of a composite product is either:

a. composed throughout of materials of limited combustibility; or

b. a Class 1 material which has a fire propagation index (I) of not more than 12 and sub-index (i_i) of not more than 6.

[†]   The classes of reaction to fire performance of A2, B, C, D and E are accompanied by additional classifications related to the production of smoke (s1, s2, s3) and/or flaming droplets/particles (d0, d1, d2).

Note: Class 0 is not a classification identified in any British Standard test.

14 Composite products defined as materials of limited combustibility (see paragraph 9 above and Table A7) should in addition comply with the test requirement appropriate to any surface rating specified in the guidance on requirements B2, B3 and B4.

15 The notional performance ratings of certain widely used generic materials or products are listed in Table A8 in terms of their performance in the traditional lining tests BS 476 Parts 6 and 7 or in accordance with BS EN 13501-1:2002, *Fire classification of construction products and building elements, Part 1-Classification using data from reaction to fire tests.*

16 Results of tests on proprietary materials are frequently given in literature available from manufacturers and trade associations.

Any reference used to substantiate the surface spread of flame rating of a material or product should be carefully checked to ensure that it is suitable, adequate and applicable to the construction to be used. Small differences in detail, such as thickness, substrate, colour, form, fixings, adhesive etc, may significantly affect the rating.

Thermoplastic materials

17 A thermoplastic material means any synthetic polymeric material which has a softening point below 200°C if tested to BS 2782 *Methods of testing plastics,* Part 1 *Thermal properties*, Method 120A: 1990 *Determination of the Vicat softening temperature of thermoplastics*. Specimens for this test may be fabricated from the original polymer where the thickness of material of the end product is less than 2.5 mm.

18 A thermoplastic material in isolation can not be assumed to protect a substrate, when used as a lining to a wall or ceiling. The surface rating of both products must therefore meet the required classification. If however, the thermoplastic material is fully bonded to a non-thermoplastic substrate, then only the surface rating of the composite will need to comply.

19 Concessions are made for thermoplastic materials used for window glazing, rooflights, and lighting diffusers within suspended ceilings, which may not comply with the criteria specified in paragraphs 11 et seq. They are described in the guidance on requirements B2 and B4.

20 For the purposes of the requirements B2 and B4 thermoplastic materials should either be used according to their classification 0- 3, under the BS 476: Parts 6 and 7 tests as described in paragraphs 11 et seq, if they have such a rating, or they may be classified TP(a) rigid, TP(a) flexible, or TP(b) according to the following methods:

TP(a) rigid:

i.   Rigid solid pvc sheet;

ii. Solid (as distinct from double- or multiple-skin) polycarbonate sheet at least 3 mm thick;

iii. Multi-skinned rigid sheet made from unplasticised pvc or polycarbonate which has a Class 1 rating when tested to BS 476: Part 7: 1971, 1987 or 1997;

iv. Any other rigid thermoplastic product, a specimen of which (at the thickness of the product as put on the market), when tested to BS 2782: 1970 as amended in 1974: Method 508A *Rate of burning (Laboratory method)*, performs so that the test flame extinguishes before the first mark, and the duration of flaming or afterglow does not exceed 5 seconds following removal of the burner.

TP(a) flexible:

Flexible products not more than 1 mm thick which comply with the Type C requirements of BS 5867 *Specification for fabrics for curtains and drapes* Part 2 *Flammability requirements* when tested to BS 5438 *Methods of test for flammability of textile fabrics when subjected to a small igniting flame applied to the face or bottom edge of vertically oriented specimens*, Test 2, 1989 with the flame applied to the surface of the specimens for 5, 15, 20 and 30 seconds respectively, but excluding the cleansing procedure; and

TP(b):

i.   Rigid solid polycarbonate sheet products less than 3 mm thick, or multiple-skin polycarbonate sheet products which do not qualify as TP(a) by test. or

ii. Other products which, when a specimen of the material between 1.5 and 3 mm thick is tested in accordance with BS 2782: 1970, as amended in 1974: Method 508A, has a rate of burning which does not exceed 50 mm/minute.

Note: If it is not possible to cut or machine a 3 mm thick specimen from the product then a 3 mm test specimen can be moulded from the same material as that used for the manufacture of the product.

Note: Currently, no new guidance is possible on the assessment or classification of thermoplastic materials under the European system since there is no generally accepted European test procedure and supporting comparative data.

Fire test methods

21 A guide to the various test methods in BS 476 and BS 2782 is given in PD 6520 Guide to fire test methods for building materials and elements of construction (available from the British Standards Institution).

A guide to the development and presentation of fire tests and their use in hazard assessment is given in BS 6336 *Guide to development and presentation of fire tests and their use in hazard assessment.*

Table A1 **Specific provisions of test for fire resistance of elements of structure etc**

Part of building	Minimum provisions when tested to the relevant part of BS 476(1) (minutes)			Minimum provisions when tested to the relevant European standard (minutes)(12)	Method of exposure
	Loadbearing capacity (2)	Integrity	Insulation		
1. **Structural** frame, beam or column	see Table A2	not applicable	not applicable	R see Table A2	exposed faces
2. **Loadbearing wall** (which is not also a wall described in any of the following items).	see Table A2	not applicable	not applicable	R see Table A2	each side separately
3. **Floors (3)**					
a. in upper storey of 2-storey dwelling house (but not over garage or basement);	30	15	15	REI 30 (9)	
b. between a shop and flat above;	60 or see Table A2 (whichever is greater)	60 or see Table A2 (whichever is greater)	60 or see Table A2 (whichever is greater)	REI 60 or see Table A2 (whichever is greater)	from underside (4)
c. any other floor, including compartment floors.	see Table A2	see Table A2	see Table A2	REI see Table A2	
4. **Roofs**					
a. any part forming an escape route;	30	30	30	REI 30	from underside (4)
b. any roof that performs the function of a floor.	see Table A2	see Table A2	see Table A2	REI see Table A2	
5. **External walls**					
a. any part less than 1000mm from any point on the relevant boundary;	See Table A2	see Table A2	see Table A2	REI see Table A2	each side separately
b. any part 1000mm or more from the relevant boundary(5);	see Table A2	see Table A2	15	REI see Table A2 (10)	from inside the building
c. any part adjacent to an external escape route (see Section 6, Diagram 22).	30	30	no provision (6)(7)	RE 30	from inside the building
6. **Compartment walls** Separating occupancies (see 9.20f)	60 or see Table A2 (whichever is less)	60 or see Table A2 (whichever is less)	60 or see Table A2 (whichever is less)	REI 60 or see Table A2 (whichever is less)	each side separately
7. **Compartment walls** (other than in item 6)	see Table A2	see Table A2	see Table A2	REI see Table A2	each side separately
8. **Protected shafts**, excluding any firefighting shaft					
a. any glazing described in Section 9, Diagram 30;	not applicable	30	no provision (7)	E 30	
b. any other part between the shaft and a protected lobby/corridor described in Diagram 30 above;	30	30	30	REI 30	each side separately
c. any part not described in (a) or (b) above.	see table A2	see table A2	see table A2	REI see table A2	
9. **Enclosure** (which does not form part of a compartment wall or a protected shaft) to a:					each side separately
a. protected stairway;	30	30	30 (8)	REI 30 (8)	
b. lift shaft	30	30	30	REI 30	

Table A1 **continued**

Part of building	Minimum provisions when tested to the relevant part of BS 476 (1) (minutes)			Minimum provisions when tested to the relevant European standard (minutes)(12)	Method of exposure
	Loadbearing capacity (2)	Integrity	Insulation		
10. **Firefighting shafts**					
a. construction separating firefighting shaft from rest of building;	120	120	120	REI 120	from side remote from shaft
	60	60	60	REI 60	from shaft side
b. construction separating firefighting stair, firefighting lift shaft and firefighting lobby.	60	60	60	REI 60	each side separately
11. **Enclosure** (which is not a compartment wall or described in item 8) to a:					each side separately
a. protected lobby;	30	30	30 (8)	REI 30 (8)	
b. protected corridor.	30	30	30 (8)	REI 30 (8)	
12. **Sub-division of a corridor**	30	30	30 (8)	REI 30 (8)	each side separately
13. **Wall separating** an attached or integral garage from a dwellinghouse	30	30	30 (8)	REI 30 (8)	from garage side
14. **Enclosure** in a flat or maisonette to a protected entrance hall, or to a protected landing.	30	30	30 (8)	REI 30 (8)	each side separately
15. **Fire-resisting construction**:					
a. in dwellings not described elsewhere;	30	30	30 (8)	REI 30 (8)	each side separately
b. enclosing places of special fire hazard (see 9.12);	30	30	30	REI 30	
c. between store rooms and sales area in shops (see 6.54)	30	30	30	REI 30	
d. fire-resisting subdivision described in Section 10, Diagram 34(b)	30	30	30	REI 30	
16. **Cavity barrier**	not applicable	30	15	EI 30 (11)	each side separately
17. **Ceiling** described in Section 10, Diagram 33 or Diagram 35	not applicable	30	30	EI 30	from underside
18. **Duct** described in paragraph 10.14e	not applicable	30	no provision	E 30	from outside
19. **Casing** around a drainage system described in Section 11, Diagram 38	not applicable	30	no provision	E 30	from outside
20. **Flue walls** described in Section 11, Diagram 39	not applicable	half the period specified in Table A2 for the compartment wall/floor	half the period specified in Table A2 for the compartment wall/floor	EI half the period specified in Table A2 for the compartment wall/floor	from outside
21. **Construction** described in Note (a) to paragraph 15.9.	not applicable	30	30	EI 30	from underside
22. **Fire doors**		see Table B1		see Table B1	

Table A1 **continued**

Notes:

1. Part 21 for loadbearing elements, Part 22 for non-loadbearing elements, Part 23 for fire-protecting suspended ceilings, and Part 24 for ventilation ducts. BS 476:Part 8 results are acceptable for items tested or assessed before 1st January 1988.
2. Applies to loadbearing elements only (see B3.ii and Appendix E).
3. Guidance on increasing the fire resistance of existing timber floors is given in BRE Digest 208 Increasing the fire resistance of existing timber floors (BRE 1988).
4. A suspended ceiling should only be relied on to contribute to the fire resistance of the floor if the ceiling meets the appropriate provisions given in Table A3.
5. The guidance in Section 14 allows such walls to contain areas which need not be fire-resisting (unprotected areas).
6. Unless needed as part of a wall in item 5a or 5b.
7. Except for any limitations on glazed elements given in Table A4.
8. See Table A4 for permitted extent of uninsulated glazed elements.
9. For the purposes of meeting the Building Regulations floors under item 3a will be deemed to have satisfied the provisions above, provided that they achieve loadbearing capacity of at least 30 minutes and integrity and insulation requirements of at least 15 minutes when tested in accordance with the relevant European test.
10. For the purposes of meeting the Building Regulations external walls under item 5b will be deemed to have satisfied the provisions above, provided that they achieve the loadbearing capacity and integrity requirements as defined in Table A2 and an insulation requirement of at least 15 minutes.
11. For the purposes of meeting the Building Regulations cavity barriers will be deemed to have satisfied the provisions above, provided that they achieve an integrity requirement of at least 30 minutes and an insulation requirement of at least 15 minutes.
12. The National classifications do not automatically equate with the equivalent classifications in the European column, therefore products cannot typically assume a European class unless they have been tested accordingly.
    "R" is the European classification of the resistance to fire performance in respect of loadbearing capacity;
    "E" is the European classification of the resistance to fire performance in respect of integrity; and
    "I" is the European classification of the resistance to fire performance in respect of insulation.

Table A2 Minimum periods of fire resistance

Purpose group of building	Minimum periods (minutes) for elements of structure in a:					
	Basement storey ($) including floor over		Ground or upper storey			
	Depth (m) of a lowest basement		Height (m) of top floor above ground, in a building or separated part of a building			
	more than 10	not more than 10	not more than 5	not more than 18	not more than 30	more than 30
1. Residential (domestic):						
a. flats and maisonettes	90	60	30*	60**†	90**	120**
b. and c. dwellinghouses	not relevant	30*	30*	60@	not relevant	not relevant
2. Residential:						
a. Institutional œ	90	60	30*	60	90	120#
b. Other residential	90	60	30*	60	90	120#
3. Office:						
- not sprinklered	90	60	30*	60	90	not permitted
- sprinklered (2)	60	60	30*	30*	60	120#
4. Shop and commercial:						
- not sprinklered	90	60	60	60	90	not permitted
- sprinklered (2)	60	60	30*	60	60	120#
5. Assembly and recreation:						
- not sprinklered	90	60	60	60	90	not permitted
- sprinklered (2)	60	60	30*	60	60	120#
6. Industrial:						
- not sprinklered	120	90	60	90	120	not permitted
- sprinklered (2)	90	60	30*	60	90	120#
7 Storage and other non-residential:						
a. any building or part not described elsewhere:						
- not sprinklered	120	90	60	90	120	not permitted
- sprinklered (2)	90	60	30*	60	90	120#
b. car park for light vehicles:						
i. open sided car park (3)	not applicable	not applicable	15* +(4)	15*+(4)	15*+(4)	60
ii. any other car park	90	60	30*	60	90	120#

Single storey buildings are subject to the periods under the heading "not more than 5". If they have basements, the basement storeys are subject to the period appropriate to their depth.

Modifications referred to in Table A2: [for application of the table see next page]

$ The floor over a basement (or if there is more than 1 basement, the floor over the topmost basement) should meet the provisions for the ground and upper storeys if that period is higher.

*   Increased to a minimum of 60 minutes for compartment walls separating buildings.

** Reduced to 30 minutes for any floor within a maisonette, but not if the floor contributes to the support of the building.

œ Multi-storey hospitals designed in accordance with the NHS Firecode document should have a minimum 60 minutes standard.

Reduced to 90 minutes for elements not forming part of the structural frame.

+ Increased to 30 minutes for elements protecting the means of escape.

† Refer to paragraph 8.10 regarding the acceptability of 30 minutes in flat conversions.

@ 30 minutes in the case of 3 storey dwellinghouses, increased to 60 minutes minimum for compartment walls separating buildings.

Notes:

1. Refer to Table A1 for the specific provisions of test.
2. "Sprinklered" means that the building is fitted throughout with an automatic sprinkler system meeting the relevant recommendations of BS 5306 Fire extinguishing installations and equipment on premises. Part 2 Specification for sprinkler systems; ie the relevant occupancy rating together with the additional requirements for life safety.
3. The car park should comply with the relevant provisions in the guidance on requirement B3, Section 12.
4. For the purposes of meeting the Building Regulations, the following types of steel elements are deemed to have satisfied the minimum period of fire resistance of 15 minutes when tested to the European test method;
   i) Beams supporting concrete floors, maximum Hp/A=230m-1 operating under full design load.
   ii) Free standing columns, maximum Hp/A=180m-1 operating under full design load.
   iii) Wind bracing and struts, maximum Hp/A=210m-1 operating under full design load.

Guidance is also available in BS5950 Structural use of steelwork in building. Part 8 Code of practice for fire resistant design.

Application of the fire resistance standards in Table A2:

a. Where one element of structure supports or carries or gives stability to another, the fire resistance of the supporting element should be no less than the minimum period of fire resistance for the other element (whether that other element is loadbearing or not).

There are circumstances where it may be reasonable to vary this principle, for example:

i. where the supporting structure is in the open air, and is not likely to be affected by the fire in the building; or

ii. the supporting structure is in a different compartment, with a fire-separating element (which has the higher standard of fire resistance) between the supporting and the separated structure; or

iii. where a plant room on the roof needs a higher fire resistance than the elements of structure supporting it.

b. Where an element of structure forms part of more than one building or compartment, that element should be constructed to the standard of the greater of the relevant provisions.

c. Where one side of a basement is (due to the slope of the ground) open at ground level, giving an opportunity for smoke venting and access for fire fighting, it may be appropriate to adopt the standard of fire resistance applicable to above-ground structures for elements of structure in that storey.

d. Although most elements of structure in a single storey building may not need fire resistance (see the guidance on requirement B3, paragraph 8.4(a)), fire resistance will be needed if the element:

i. is part of (or supports) an external wall and there is provision in the guidance on requirement B4 to limit the extent of openings and other unprotected areas in the wall; or

ii. is part of (or supports) a compartment wall, including a wall common to two or more buildings, or a wall between a dwelling house and an attached or integral garage; or

iii. supports a gallery.

For the purposes of this paragraph, the ground storey of a building which has one or more basement storeys and no upper storeys, may be considered as a single storey building. The fire resistance of the basement storeys should be that appropriate to basements.

Table A3 **Limitations on fire-protecting suspended ceilings** (see Table A1, Note 4)

Height of building or separated part (m)	Type of floor	Provision for fire resistance or floor (minutes)	Description of suspended ceiling
less than 18	not compartment	60 or less	Type W, X, Y or Z
	compartment	less than 60	
		60	Type X, Y or Z
18 or more	any	60 or less	Type Y or Z
no limit	any	more than 60	Type Z

Notes:

1. Ceiling type and description (the change from Types A-D to Types W-Z is to avoid confusion with Classes A-D (European)):
    W. Surface of ceiling exposed to the cavity should be Class 0 or Class 1 (National) or Class C-s3, d2 or better (European).
    X. Surface of ceiling exposed to the cavity should be Class 0 (National) or Class B-s3, d2 or better (European).
    Y. Surface of ceiling exposed to the cavity should be Class 0 (National) or Class B-s3, d2 or better (European). Ceiling should not contain easily openable access panels.
    Z. Ceiling should be of a material of limited combustibility (National) or of Class A2-s3, d2 or better (European) and not contain easily openable access panels. Any insulation above the ceiling should be of a material of limited combustibility (National) or Class A2-s3, d2 or better (European).
2. Any access panels provided in fire protecting suspended ceilings of type Y or Z should be secured in position by releasing devices or screw fixings, and they should be shown to have been tested in the ceiling assembly in which they are incorporated.

3. European classifications

The National classifications do not automatically equate with the equivalent European classifications, therefore products cannot typically assume a European class unless they have been tested accordingly.
When a classification includes "s3, d2", this means that there is no limit set for smoke production and/or flaming droplets/particles.

Table A4 Limitations on the use of uninsulated glazed elements on escape routes

(These limitations do not apply to glazed elements which satisfy the relevant insulation criterion, see Table A1) (See BS 5588: Part 7 for glazing to atria; see BS 5588: Part 8 for glazing to refuges)

Position of glazed element	Maximum total glazed area in parts of a building with access to:			
	a single stairway		more than one stairway	
	walls	door leaf	walls	door leaf
Single family dwellinghouses 1. a. Within the enclosures of: i. a protected stairway, or within fire-resisting separation shown in Section 2 Diagram 3; or	fixed fanlights only	unlimited	fixed fanlights only	unlimited
ii. an existing stair (see para 2.18).	unlimited	unlimited	unlimited	unlimited
b. Within fire-resisting separation: i. shown in Section 2 Diagram 4, or ii. described in paras 2.13b & 2.20.	unlimited above 100 mm from floor	unlimited above 100 mm from floor	unlimited above 100 mm from floor	unlimited above 100 mm from floor
c. Existing window between an attached/integral garage and the house.	unlimited	not applicable	unlimited	not applicable
Flats and maisonettes 2. Within the enclosures of a protected entrance hall or protected landing or within fire-resisting separation shown in Section 3 Diagram 9.	fixed fanlights only	unlimited above 1100mm from floor	fixed fanlights only	unlimited above 1100 mm from floor
General (except dwellinghouses) 3. Between residential/sleeping accommodation and a common escape route (corridor, lobby or stair).	nil	nil	nil	nil
4. Between a protected stairway (1) and: a. the accommodation; or b. a corridor which is not a protected corridor. Other than in item 3 above.	nil	25% of door area	unlimited above 1100 mm [2]	50% of door area
5. Between: a. a protected stairway (1) and a protected lobby or protected corridor; or b. accommodation and a protected lobby. Other than in item 3 above.	unlimited above 1100 mm from floor	unlimited above 100 mm from floor	unlimited above 100 mm from floor	unlimited above 100 mm from floor
6. Between the accommodation and a protected corridor forming a dead end. Other than in item 3 above.	unlimited above 1100 mm from floor	unlimited above 100 mm from floor	unlimited above 1100 mm from floor	unlimited above 100 mm from floor
7. Between accommodation and any other corridor; or subdividing corridors. Other than in item 3 above.	not applicable	not applicable	unlimited above 100 mm from floor	unlimited above 100 mm from floor
8. Adjacent an external escape route described in para 4.27.	unlimited above 1100 mm from paving	unlimited above 1100 mm from paving	unlimited above 1100 mm from paving	unlimited above 1100 mm from paving
9. Adjacent an external escape stair (see para 6.25 & Diagram 22) or roof escape (see para 6.35).	unlimited	unlimited	unlimited	unlimited

Notes:
1. If the protected stairway is also a protected shaft (see paragraph 9.36) or a firefighting stair (see Section 18) there may be further restrictions on the uses of glazed elements.
2. Measured vertically from the landing floor level or the stair pitch line.
3. The 100 mm limit is intended to reduce the risk of fire spread from a floor covering.
4. Items 1c, 3 and 6 apply also to single storey buildings.

Table A5 Notional designations of roof coverings

Part i: Pitched roofs covered with slates or tiles

Covering material	Supporting structure	Designation
1. Natural slates 2. Fibre reinforced cement slates 3. Clay tiles 4. Concrete tiles	timber rafters with or without underfelt, sarking, boarding, woodwool slabs, compressed straw slabs, plywood, wood chipboard, or fibre insulating board	AA

Note: Although the Table does not include guidance for roofs covered with bitumen felt, it should be noted that there is a wide range of materials on the market and information on specific products is readily available from manufacturers.

Part ii: Pitched roofs covered with self-supporting sheet

Roof covering material	Construction	Supporting structure	Designation
1. Profiled sheet of galvanised steel, aluminium, fibre reinforced cement, or pre-painted (coil coated) steel or aluminium with a pvc or pvf2 coating	single skin without underlay, or with underlay or plasterboard, fibre insulating board, or woodwool slab	structure of timber, steel or concrete	AA
2. Profiled sheet of galvanised steel, aluminium, fibre reinforced cement, or pre-painted (coil coated) steel or aluminium with a pvc or pvf2 coating	double skin without interlayer, or with interlayer of resin bonded glass fibre, mineral wool slab, polystyrene, or polyurethane	structure of timber, steel or concrete	AA

Part iii. Flat roofs covered with bitumen felt

A flat roof comprising of bitumen felt should (irrespective of the felt specification) be deemed to be of designation AA if the felt is laid on a deck constructed of 6 mm plywood, 12.5 mm wood chipboard, 16 mm (finished) plain edged timber boarding, compressed straw slab, screeded wood wool slab, profiled fibre reinforced cement or steel deck (single or double skin) with or without fibre insulating board overlay, profiled aluminium deck (single or double skin) with or without fibre insulating board overlay, or concrete or clay pot slab (insitu or pre cast), and has a surface finish of:
a. bitumen-bedded stone chippings covering the whole surface to a depth of at least 12.5 mm;
b. bitumen-bedded tiles of a non-combustible material;
c. sand and cement screed; or
d. macadam.

Part iv. Pitched or flat roofs covered with fully supported material

Covering material	Supporting structure	Designation
1. Aluminium sheet 2. Copper sheet 3. Zinc sheet 4. Lead sheet 5. Mastic asphalt 6. Vitreous enamelled steel 7. Lead/tin alloy coated steel sheet 8. Zinc/aluminium alloy coated steel sheet 9. Pre-painted (coil coated) steel sheet including liquid-applied pvc coatings	timber joists and: tongued and grooved boarding, or plain edged boarding	AA*
	steel or timber joists with deck of: woodwool slabs, compressed straw slab, wood chipboard, fibre insulating board, or 9.5 mm plywood	AA
	concrete or clay pot slab (insitu or pre-cast) or non-combustible deck of steel, aluminium, or fibre cement (with or without insulation)	AA

Notes:
*   Lead sheet supported by timber joists and plain edged boarding should be regarded as having a BA designation.

Table A6 **Use and definitions of non-combustible materials**

References in AD.B guidance to situations where such materials should be used	Definitions of non-combustible materials	
	National class	**European class**
1. ladders referred to in the guidance to B1, paragraph 6.22. 2. refuse chutes meeting the provisions in the guidance to B3, paragraph 9.35c. 3. suspended ceilings and their supports where there is provision in the guidance to B3, paragraph 10.13, for them to be constructed of non-combustible materials. 4. pipes meeting the provisions in the guidance to B3, Table 15. 5. flue walls meeting the provisions in the guidance to B3, Diagram 39. 6. construction forming car parks referred to in the guidance to B3, paragraph 12.3.	a. Any material which when tested to BS476 : Part 11 does not flame nor cause any rise in temperature on either the centre (specimen) or furnace thermocouples b. Totally inorganic materials such as concrete, fired clay, ceramics, metals, plaster and masonry containing not more than 1% by weight or volume of organic material. (Use in buildings of combustible metals such as magnesium/aluminium alloys should be assessed in each individual case. c. Concrete bricks or blocks meeting BS 6073 : Part 1 d. Products classified as non-combustible under BS 476 : Part 4	a. Any material classified as class A1 in accordance with BS EN 13501-1:2002, Fire classification of construction products and building elements, Part 1-Classification using data from reaction to fire tests. b. Products made from one or more of the materials considered as Class A1 without the need for testing, as defined in Commission Decision 96/603/EC of 4th October 1996 establishing the list of products belonging to Class A1 "No contribution to fire" provided for in the Decision 94/611/EC implementing Article 20 of the Council Directive 89/106/EEC on construction products. None of the materials shall contain more than 1.0% by weight or volume (whichever is the lower) of homogeneously distributed organic material.
		Note: The National classifications do not automatically equate with the equivalent classifications in the European column, therefore products cannot typically assume a European class unless they have been tested accordingly.

Table A7 **Use and definitions of materials of limited combustibility**

References in AD.B guidance to situations where such materials should be used	Definitions of non-combustible materials	
	National class	**European class**

1. stairs where there is provision in the guidance to B1 for them to be constructed of materials of limited combustibility (see 6.19).
 	
2. materials above a suspended ceiling meeting the provisions in the guidance to B3, paragraph 10.13.
 	
3. reinforcement/support for fire-stopping referred to in the guidance to B3, see 11.13.
 	
4. roof coverings meeting provisions:
   a. in the guidance to B3, paragraph 10.11 or
   b. in the guidance to B4, Table 17 or
   c. in the guidance to B4, Diagram 47.
 	
5. roof deck meeting the provisions of the guidance to B3, Diagram 28a.
 	
6. class 0 materials meeting the provisions in Appendix A, paragraph 13(a).
 	
7. ceiling tiles or panels of any fire protecting suspended ceiling (Type Z) in Table A3.
 	
8. compartment walls and compartment floors in hospitals referred to in paragraph 9.32.
	a. Any non-combustible material listed in Table A6.
	
	b. Any material of density 300/kg/m^3 or more, which when tested to BS476: Part 11, does not flame and the rise in temperature on the furnace thermocouple is not more than 20°C.
	
	c. Any material with a non-combustible core at least 8mm thick having combustible facings (on one or both sides) not more than 0.5mm thick. (Where a flame spread rating is specified, these materials must also meet the appropriate test requirements).
		a. Any material listed in Table A6.
		
		b. Any material/product classified as Class A2-s3, d2 or better in accordance with BS EN 13501-1:2002, *Fire classification of construction products and building elements, Part 1-Classification using data from reaction to fire tests.*

9. insulation material in external wall construction referred to in paragraph 13.7. 10. insulation above any fire-protecting suspended ceiling (Type Z) in Table A3.	Any of the materials (a), (b) or (c) above, or: d. Any material of density less than 300kg/m^3, which when tested to BS476:Part 11, does not flame for more than 10 seconds and the rise in temperature on the centre (specimen) thermocouple is not more than 35°C and on the furnace thermocouple is not more than 25°C.	Any of the materials/products (a) or (b) above
		Notes: 1. The National classifications do not automatically equate with the equivalent classifications in the European column, therefore products cannot typically assume a European class unless they have been tested accordingly. 2. When a classification includes "s3, d2", this means that there is no limit set for smoke production and/or flaming droplets/particles.

Table A8 Typical performance ratings of some generic materials and products

Rating	Material or product
Class 0 (National)	1. any non-combustible material or material of limited combustibility. (Composite products listed in Table A7 must meet test requirements given in Appendix A, paragraph 13(b)).
	2. brickwork, blockwork, concrete and ceramic tiles.
	3. plasterboard (painted or not with a PVC facing not more than 0.5mm thick) with or without an air gap or fibrous or cellular insulating material behind.
	4. woodwool cement slabs.
	5. mineral fibre tiles or sheets with cement or resin binding.
Class 3 (National)	6. timber or plywood with a density more than 400kg/m^3, painted or unpainted.
	7. wood particle board or hardboard, either untreated or painted.
	8. standard glass reinforced polyesters.
Class A1 (European)	9. Any material that achieves this class and is defined as "classified without further test" in a published Commission Decision.
Class A2-s3, d2 (European)	10. Any material that achieves this class and is defined as "classified without further test" in a published Commission Decision.
Class B-s3, d2 (European)	11. Any material that achieves this class and is defined as "classified without further test" in a published Commission Decision.
Class C-s3, d2 (European)	12. Any material that achieves this class and is defined as "classified without further test" in a published Commission Decision.
Class D-s3, d2 (European)	13. Any material that achieves this class and is defined as "classified without further test" in a published Commission Decision.

Notes (National):
1. Materials and products listed under Class 0 also meet Class 1.
2. Timber products listed under Class 3 can be brought up to Class 1 with appropriate proprietary treatments.
3. The following materials and products may achieve the ratings listed below. However, as the properties of different products with the same generic description vary, the ratings of these materials/ products should be substantiated by test evidence.
   Class 0 – aluminium faced fibre insulating board, flame retardant decorative laminates on a calcium silicate board, thick polycarbonate sheet, phenolic sheet and UPVC;
   Class 1 – phenolic or melamine laminates on a calcium silicate substrate and flame retardant decorative laminates on a combustible substrate.

Notes (European):
For the purposes of the Building Regulations
1. Materials and products listed under Class A1 also meet Classes A2-s3, d2, B-s3, d2, C-s3, d2 and D-s3, d2.
2. Materials and products listed under Class A2-s3, d2 also meet Classes B-s3, d2, C-s3, d2 and D-s3, d2.
3. Materials and products listed under Class B-s3, d2 also meet Classes C-s3, d2 and D-s3, d2.
4. Materials and products listed under Class C-s3, d2 also meet Class D-s3, d2.
5. The performance of timber products listed under Class D-s3, d2 can be improved with appropriate proprietary treatments.
6. Materials covered by the CWFT process (classification without further testing) can be found by accessing the European Commission's website via the link on the ODPM's web site www.odpm.gov.uk/bregs/cpd/index.htm.
7. The National classifications do not automatically equate with the equivalent classifications in the European column, therefore products cannot typically assume a European class unless they have been tested accordingly.
8. When a classification includes "s3, d2", this means that there is no limit set for smoke production and/or flaming droplets/particles.

Appendix B

FIRE DOORS

1 All fire doors should have the appropriate performance given in Table B1 either:

a. by their performance under test to BS 476 *Fire tests on building materials and structures, Part 22 Methods for determination of the fire resistance of non-loadbearing elements of construction*, in terms of integrity for a period of minutes, eg FD30. A suffix (S) is added for doors where restricted smoke leakage at ambient temperatures is needed; or

b. as determined with reference to Commission Decision 2000/367/EC of 3rd May 2000 implementing Council Directive 89/106/EEC as regards the classification of the resistance to fire performance of construction products, construction works and parts thereof. All fire doors should be classified in accordance with BS EN 13501-2: xxxx, *Fire classification of construction products and building elements, Part 2 – Classification using data from fire resistance tests (excluding products for use in ventilation systems)*. They are tested to the relevant European method from the following:

BS EN 1634-1: 2000, *Fire resistance tests for door and shutter assemblies, Part 1 – Fire doors and shutters;*

BS EN 1634-2: xxxx, *Fire resistance tests for door and shutter assemblies, Part 2 – Fire door hardware;*

BS EN 1634- 3: xxxx, *Fire resistance tests for door and shutter assemblies, Part 3 – Smoke control doors.*

The performance requirement is in terms of integrity (E) for a period of minutes. An additional classification of S_a is used for all doors where restricted smoke leakage at ambient temperatures is needed.

The requirement (in either case) is for test exposure from each side of the door separately, except in the case of lift doors which are tested from the landing side only.

Notes:

1. The designation of xxxx is used for standards that are not yet published. The latest version of any standard may be used provided that it continues to address the relevant requirements of the Regulations.

2. Until such time that the relevant harmonised product standards are published, for the purposes of meeting the Building Regulations, products tested in accordance with BS EN 1634-1 (with or without pre-fire test mechanical conditioning) will be deemed to have satisfied the provisions provided that they achieve the minimum fire resistance in terms of integrity, as detailed in Table B1.

2 All fire doors should be fitted with an automatic self-closing device except for fire doors to cupboards and to service ducts which are normally kept locked shut.

Note: All rolling shutters should be capable of being opened and closed manually for firefighting purposes (see Section 18, paragraph 18.13).

3 Where a self-closing device would be considered a hindrance to the normal approved use of the building, self-closing fire doors may be held open by:

a. a fusible link (but not if the door is fitted in an opening provided as a means of escape unless it complies with paragraph 4 below); or

b. an automatic release mechanism actuated by an automatic fire detection & alarm system; or

c. a door closer delay device.

4 Two fire doors may be fitted in the same opening so that the total fire resistance is the sum of their individual fire resistances, provided that each door is capable of closing the opening. In such a case, if the opening is provided as a means of escape, both doors should be self-closing, but one of them may be fitted with an automatic self-closing device and be held open by a fusible link if the other door is capable of being easily opened by hand and has at least 30 minutes fire resistance.

5 Because fire doors often do not provide any significant insulation, there should be some limitation on the proportion of doorway openings in compartment walls. Therefore no more than 25% of the length of a compartment wall should consist of door openings, unless the doors provide both integrity and insulation to the appropriate level (see Appendix A, Table A2).

6 Roller shutters across a means of escape should only be released by a heat sensor, such as a fusible link or electric heat detector, in the immediate vicinity of the door. Closure of shutters in such locations should not be initiated by smoke detectors or a fire alarm system, **unless** the shutter is also intended to partially descend to form part of a boundary to a smoke reservoir.

7 Unless shown to be satisfactory when tested as part of a fire door assembly, the essential components of any hinge on which a fire door is hung should be made entirely from materials having a melting point of at least 800°C.

8 Except for doors identified in paragraph 9 below, all fire doors should be marked with the appropriate fire safety sign complying with BS 5499 *Fire safety signs, notices and graphic symbols*, Part 1 *Specification for fire safety signs*, according to whether the door is:

a. to be kept closed when not in use;

b. to be kept locked when not in use; or

c. held open by an automatic release mechanism.

Fire doors to cupboards and to service ducts should be marked on the outside; all other fire doors on both sides.

9 The following fire doors are not required to comply with paragraph 8 above:

a. doors within dwelling houses;

b. doors to and within flats or maisonettes;

c. bedroom doors in 'Other-residential' premises; and

d. lift entrance/landing doors.

10 Tables A1 and A2 set out the minimum periods of fire resistance for the elements of structure to which performance of some doors is linked. Table A4 sets out limitations on the use of uninsulated glazing in fire doors.

11 BS 8214 *Code of practice for fire door assemblies with non-metallic leaves* gives recommendations for the specification, design, construction, installation and maintenance of fire doors constructed with non-metallic door leaves.

Guidance on timber fire-resisting doorsets, in relation to the new European test standard, may be found in *"Timber Fire-Resisting Doorsets: maintaining performance under the new European test standard"* published by TRADA.

Guidance for metal doors is given in Code of *practice for fire-resisting metal doorsets* published by the DSMA (Door and Shutter Manufacturers' Association) in 1999.

12 Hardware used on fire doors can significantly affect performance in fire. Notwithstanding the guidance in this Approved Document guidance is available in *"Hardware for timber and escape doors"* published by the Builders Hardware Industry Federation in November 2000.

Table B1 **Provisions for fire doors**

Position of door	Minimum fire resistance of door in terms of integrity (minutes) when tested to BS 476 part 22(1)	Minimum fire resistance of door in terms of integrity (minutes) when tested to the relevant European standard (3)
1. In a compartment wall separating buildings	As for the wall in which the door is fitted, but a minimum of 60	As for the wall in which the door is fitted, but a minimum of 60
2. In a compartment wall: a. If it separates a flat or maisonette from a space in common use;	FD 30S (2)	E30 S$_a$ (2)
b. Enclosing a protected shaft forming a stairway situated wholly or partly above the adjoining ground in a building used for Flats, Other Residential, Assembly and Recreation, or Office purposes;	FD 30S (2)	E30 S$_a$ (2)
c. enclosing a protected shaft forming a stairway not described in (b) above;	Half the period of fire resistance of the wall in which it is fitted, but 30 minimum and with suffix S (2)	Half the period of fire resistance of the wall in which it is fitted, but 30 minimum and with suffix S$_a$(2)
d. enclosing a protected shaft forming a lift or service shaft;	Half the period of fire resistance of the wall in which it is fitted, but 30 minimum	Half the period of fire resistance of the wall in which it is fitted, but 30 minimum
e. not described in (a), (b), (c) or (d) above.	As for the wall it is fitted in, but add S (2) if the door is used for progressive horizontal evacuation under the guidance to B1	As for the wall it is fitted in, but add S$_a$(2) if the door is used for progressive horizontal evacuation under the guidance to B1
3. In a compartment floor	As for the floor in which it is fitted	As for the floor in which it is fitted
4. Forming part of the enclosures of: a. a protected stairway (except where described in item 9); or	FD 30S (2)	E30 S$_a$ (2)
b. a lift shaft (see paragraph 6.42b); which does not form a protected shaft in 2(b), (c) or (d) above.	FD 30	E30
5. Forming part of the enclosure of: a. a protected lobby approach (or protected corridor) to a stairway;	FD 30S (2)	E30 S$_a$ (2)
b. any other protected corridor; or	FD 20S (2)	E20 S$_a$ (2)
c. a protected lobby approach to a lift shaft (see paragraph 6.42)	FD 30S (2)	E30 S$_a$ (2)
6. Affording access to an external escape route	FD 30	E30
7. Sub-dividing: a. corridors connecting alternative exits;	FD 20S (2)	E20 S$_a$ (2)
b. dead-end portions of corridors from the remainder of the corridor	FD 20S (2)	E20 S$_a$ (2)
8. Any door: a. within a cavity barrier;	FD 30	E30
b. between a dwellinghouse and a garage	FD 30	E30

Table B1 **continued**

Position of door	Minimum fire resistance of door in terms of integrity (minutes) when tested to BS 476 part 22(1)	Minimum fire resistance of door in terms of integrity (minutes) when tested to the relevant European standard
9. Any door: a. forming part of the enclosures to a protected stairway in a single family dwelling house;	FD 20	E20
b. forming part of the enclosure to a protected entrance hall or protected landing in a flat or maisonette;	FD 20	E20
c. within any other fire-resisting construction in a dwelling not described elsewhere in this table	FD 20	E20

Notes:

1. To BS 476: Part 22 (or BS 476: Part 8 subject to paragraph 5 in Appendix A).

2. Unless pressurization techniques complying with BS 5588: Part 4 *Fire precautions in the design, construction and use of buildings, Code of practice for smoke control using pressure differentials* are used, these doors should also either:

   (a) have a leakage rate not exceeding $3m^3$/m/hour (head and jambs only) when tested at 25 Pa under BS 476 *Fire tests on building materials and structures*, Section 31.1 *Methods for measuring smoke penetration through doorsets and shutter assemblies, Method of measurement under ambient temperature conditions*; or

   (b) meet the additional classification requirement of S_a when tested to BS EN 1634- 3:xxxx, *Fire resistance tests for door and shutter assemblies, Part 3- Smoke control doors.*

3. The National classifications do not automatically equate with the equivalent classifications in the European column, therefore products cannot typically assume a European class unless they have been tested accordingly.

Appendix C

METHODS OF MEASUREMENT

1 Some form of measurement is an integral part of many of the provisions in this document. Diagrams C1 to C5 show how the various forms of measurement should be made.

Note: See Approved Document B1, paragraph B1.xxv for methods of measurement of occupant capacity, travel distance and width of doors, escape routes and stairs, which are specific to means of escape in case of fire.

Diagram C1 **Cubic capacity**

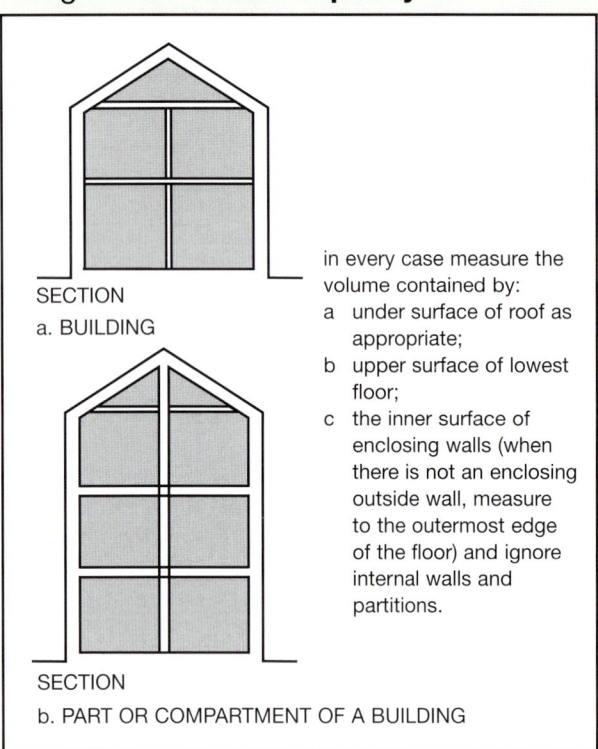

SECTION
a. BUILDING

SECTION
b. PART OR COMPARTMENT OF A BUILDING

in every case measure the volume contained by:
a under surface of roof as appropriate;
b upper surface of lowest floor;
c the inner surface of enclosing walls (when there is not an enclosing outside wall, measure to the outermost edge of the floor) and ignore internal walls and partitions.

Diagram C2 **Area**

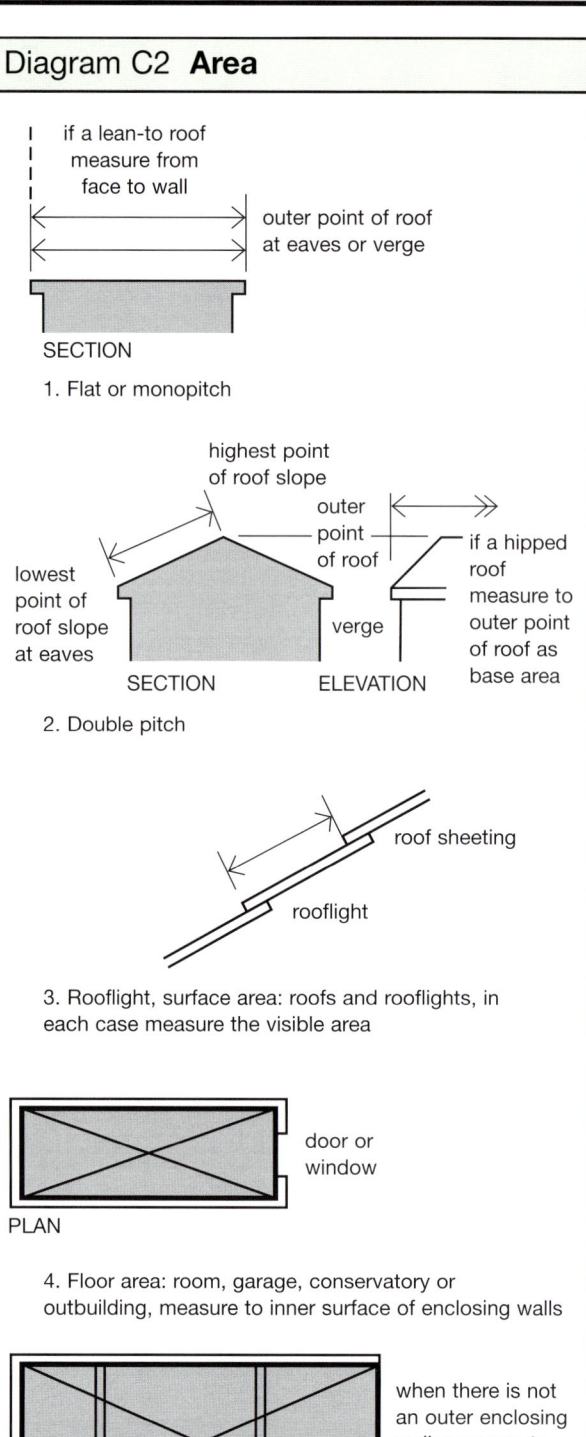

if a lean-to roof measure from face to wall

outer point of roof at eaves or verge

SECTION

1. Flat or monopitch

highest point of roof slope

outer point of roof

lowest point of roof slope at eaves

verge

if a hipped roof measure to outer point of roof as base area

SECTION   ELEVATION

2. Double pitch

roof sheeting

rooflight

3. Rooflight, surface area: roofs and rooflights, in each case measure the visible area

door or window

PLAN

4. Floor area: room, garage, conservatory or outbuilding, measure to inner surface of enclosing walls

when there is not an outer enclosing wall, measure to the outmost edge of the floor slab

PLAN

5. Floor area: storey, part or compartment, measure to inner surface of enclosing walls and include internal walls and partitions

Diagram C3 Height of building

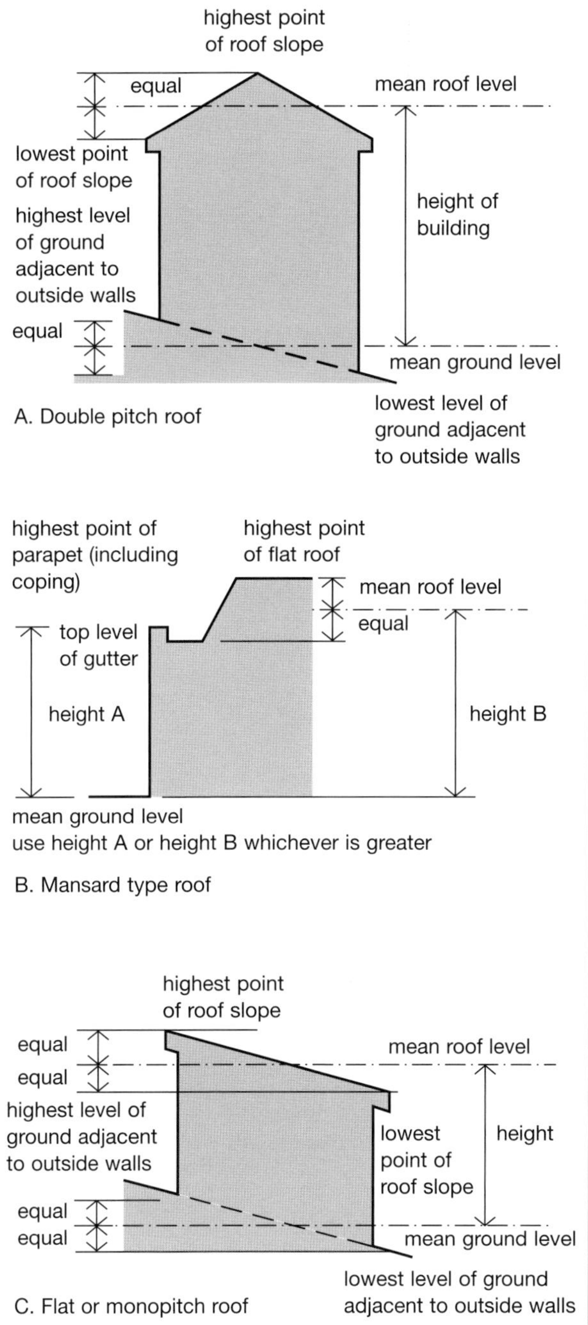

highest point
of roof slope
equal
mean roof level
lowest point
of roof slope
highest level
of ground
adjacent to
outside walls
height of
building
equal
mean ground level
lowest level of
ground adjacent
to outside walls
A. Double pitch roof
highest point of
parapet (including
coping)
highest point
of flat roof
mean roof level
equal
top level
of gutter
height A
height B
mean ground level
use height A or height B whichever is greater
B. Mansard type roof
highest point
of roof slope
equal
equal
mean roof level
highest level of
ground adjacent
to outside walls
lowest
point of
roof slope
height
equal
equal
mean ground level
lowest level of ground
adjacent to outside walls
C. Flat or monopitch roof

Diagram C4 Number of storeys

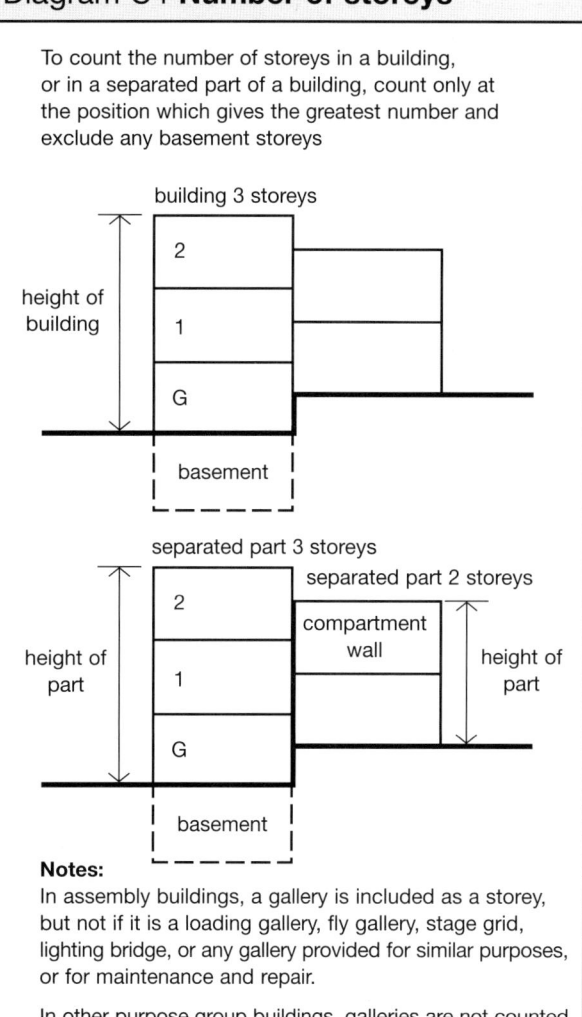

To count the number of storeys in a building, or in a separated part of a building, count only at the position which gives the greatest number and exclude any basement storeys

building 3 storeys
height of
building
2
1
G
basement
separated part 3 storeys
separated part 2 storeys
compartment
wall
height of
part
2
1
G
height of
part
basement

Notes:
In assembly buildings, a gallery is included as a storey, but not if it is a loading gallery, fly gallery, stage grid, lighting bridge, or any gallery provided for similar purposes, or for maintenance and repair.

In other purpose group buildings, galleries are not counted as a storey.

Diagram C5 Height of top storey in building

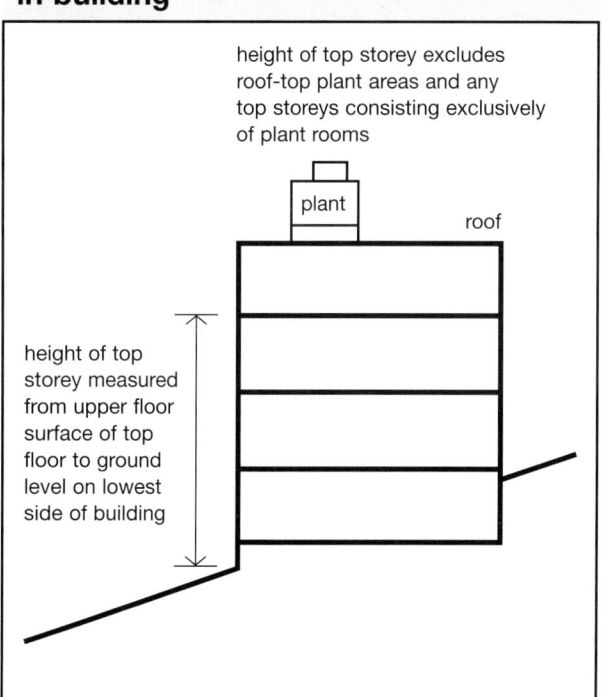

height of top storey excludes
roof-top plant areas and any
top storeys consisting exclusively
of plant rooms
plant
roof
height of top
storey measured
from upper floor
surface of top
floor to ground
level on lowest
side of building

Appendix D

PURPOSE GROUPS

1 Many of the provisions in this document are related to the use of the building. The use classifications are termed purpose groups and represent different levels of hazard. They can apply to a whole building, or (where a building is compartmented) to a compartment in the building, and the relevant purpose group should be taken from the main use of the building or compartment.

2 Table D1 sets out the purpose group classification.

Note: It is only of relevance to this Approved Document.

Ancillary and main uses

3 In some situations there may be more than one use involved in a building or compartment, and in certain circumstances it is appropriate to treat the different use as belonging to a purpose group in its own right. These situations are:

a. where the ancillary use is a flat or maisonette; or

b. where the building or compartment is more than 280m^2 in area and the ancillary use is of an area that is more than a fifth of the total floor area of the building or compartment; or

c. storage in a building or compartment of purpose group 4 (shop or commercial), where the storage amounts to more than 1/3rd of the total floor area of the building or compartment and the building or compartment is more than 280m^2 in area.

4 Some buildings may have two or more main uses that are not ancillary to one another. For example offices over shops from which they are independent. In such cases, each of the uses should be considered as belonging to a purpose group in its own right.

5 In other cases, and particularly in some large buildings, there may be a complex mix of uses. In such cases it is necessary to consider the possible risk that one part of a complex may have on another and special measures to reduce the risk may be necessary.

Table D1 Classification of purpose groups

Title	Group	Purpose for which the building or compartment of a building is intended to be used
Residential* (dwellings)	1(a)	Flat or maisonette.
	1(b)	Dwellinghouse which contains a habitable storey with a floor level which is more than 4.5m above ground level.
	1(c)	Dwellinghouse which does not contain a habitable storey with a floor level which is more than 4.5m above ground level.
Residential (Institutional)	2(a)	Hospital, home, school or other similar establishment used as living accommodation for, or for the treatment, care or maintenance of persons suffering from disabilities due to illness or old age or other physical or mental incapacity, or under the age of five years, or place of lawful detention, where such persons sleep on the premises.
(Other)	2(b)	Hotel, boarding house, residential college, hall of residence, hostel, and any other residential purpose not described above.
Office	3	Offices or premises used for the purpose of administration, clerical work (including writing, book keeping, sorting papers, filing, typing, duplicating, machine calculating, drawing and the editorial preparation of matter for publication, police and fire service work), handling money (including banking and building society work), and communications (including postal, telegraph and radio communications) or radio, television, film, audio or video recording, or performance [not open to the public] and their control.
Shop and Commercial	4	Shops or premises used for a retail trade or business (including the sale to members of the public of food or drink for immediate consumption and retail by auction, self-selection and over-the-counter wholesale trading, the business of lending books or periodicals for gain and the business of a barber or hairdresser) and premises to which the public is invited to deliver or collect goods in connection with their hire repair or other treatment, or (except in the case of repair of motor vehicles) where they themselves may carry out such repairs or other treatments.
Assembly and Recreation	5	Place of assembly, entertainment or recreation; including bingo halls, broadcasting, recording and film studios open to the public, casinos, dance halls; entertainment, conference, exhibition and leisure centres; funfairs and amusement arcades; museums and art galleries; non-residential clubs, theatres, cinemas and concert halls; educational establishments, dancing schools, gymnasia, swimming pool buildings, riding schools, skating rinks, sports pavilions, sports stadia; law courts; churches and other buildings of worship, crematoria; libraries open to the public, non-residential day centres, clinics, health centres and surgeries; passenger stations and termini for air, rail, road or sea travel; public toilets; zoos and menageries.
Industrial	6	Factories and other premises used for manufacturing, altering, repairing, cleaning, washing, breaking-up, adapting or processing any article; generating power or slaughtering livestock.
Storage and other non-residential+	7(a)	Place for the storage or deposit of goods or materials [other than described under 7(b)] and any building not within any of the purpose groups 1 to 6.
	7(b)	Car parks designed to admit and accommodate only cars, motorcycles and passenger or light goods vehicles weighing no more than 2500 kg gross.

Notes:

This table only applies to Part B.

* Includes any surgeries, consulting rooms, offices or other accommodation, not exceeding 50m² in total, forming part of a dwelling and used by an occupant of the dwelling in a professional or business capacity.

+ A detached garage not more than 40m² in area is included in purpose group 1(c); as is a detached open carport of not more than 40m², or a detached building which consists of a garage and open carport where neither the garage nor open carport exceeds 40m² in area.

Appendix E

DEFINITIONS

Note: Except for the items marked * (which are from the Building Regulations), these definitions apply only to Part B.

Access room A room through which passes the only escape route from an inner room.

Accommodation stair A stair, additional to that or those required for escape purposes, provided for the convenience of occupants.

Alternative escape routes Escape routes sufficiently separated by either direction and space, or by fire-resisting construction, to ensure that one is still available should the other be affected by fire.

> **Note:** A second stair, balcony or flat roof which enables a person to reach a place free from danger from fire, is considered an alternative escape route for the purposes of a dwelling house.

Alternative exit One of two or more exits, each of which is separate from the other.

Appliance ventilation duct A duct provided to convey combustion air to a gas appliance.

Atrium (plural atria) A space within a building, not necessarily vertically aligned, passing through one or more structural floors.

> **Note:** Enclosed lift wells, enclosed escalator wells, building services ducts and stairways are not classified as atria.

Automatic release mechanism A device which will allow a door held open by it to close automatically in the event of each or any one of the following:

a. detection of smoke by automatic apparatus suitable in nature, quality and location;

b. operation of a hand operated switch fitted in a suitable position;

c. failure of electricity supply to the device, apparatus or switch;

d. operation of the fire alarm system if any.

Automatic self-closing device A device which is capable of closing the door from any angle and against any latch fitted to the door.

> **Note:** Rising butt hinges which do not meet the above criteria are acceptable where the door is:
>
> a. to (or within) a dwelling;
>
> b. between a dwelling house and its garage; or
>
> c. in a cavity barrier.

Basement storey A storey with a floor which at some point is more than 1200mm below the highest level of ground adjacent to the outside walls. (However, see Appendix A, Table A2, for situations where the storey is considered to be a basement only because of a sloping site).

Boundary The boundary of the land belonging to the building, or where the land abuts a road, railway, canal or river, the centreline of that road, railway, canal or river. (See Diagram 41).

* Building Any permanent or temporary building but not any other kind of structure or erection. A reference to a building includes a reference to part of a building.

Building Control Body A term used to include both Local Authority Building Control and Approved Inspectors.

Cavity barrier A construction, other than a smoke curtain, provided to close a concealed space against penetration of smoke or flame, or provided to restrict the movement of smoke or flame within such a space.

Ceiling A part of a building which encloses and is exposed overhead in a room, protected shaft or circulation space. (The soffit of a rooflight is included as part of the surface of the ceiling, but not the frame. An upstand below a rooflight would be considered as a wall).

Circulation space A space (including a protected stairway) mainly used as a means of access between a room and an exit from the building or compartment.

Class 0 A product performance classification for wall and ceiling linings. The relevant test criteria are set out in Appendix A, paragraph 13.

Common balcony A walkway, open to the air on one or more sides, forming part of the escape route from more than one flat or maisonette.

Common stair An escape stair serving more than one flat or maisonette.

Compartment (fire) A building or part of a building, comprising one or more rooms, spaces or storeys, constructed to prevent the spread of fire to or from another part of the same building, or an adjoining building. (A roof space above the top storey of a compartment is included in that compartment). (See also "Separated part").

Compartment wall or floor A fire-resisting wall/floor used in the separation of one fire compartment from another. (Constructional provisions are given in Section 9)

Concealed space or cavity A space enclosed by elements of a building (including a suspended ceiling) or contained within an element, but not a room, cupboard, circulation space, protected shaft or space within a flue, chute, duct, pipe or conduit.

Corridor access A design of a building containing flats in which each dwelling is approached via a common horizontal internal access or circulation space which may include a common entrance hall.

Dead-end Area from which escape is possible in one direction only.

Direct distance The shortest distance from any point within the floor area, measured within the external enclosures of the building, to the nearest storey exit ignoring walls, partitions and fittings, other than the enclosing walls/ partitions to protected stairways.

Dwelling A unit of residential accommodation occupied (whether or not as a sole or main residence):

a. by a single person or by people living together as a family; or

b. by not more than 6 residents living together as a single household, including a household where care is provided for residents.

* **Dwelling-house** does not include a flat or a building containing a flat.

Element of structure

a. a member forming part of the structural frame of a building or any other beam or column;

b. a loadbearing wall or loadbearing part of a wall;

c. a floor;

d. a gallery (but not a loading gallery, fly gallery, stage grid, lighting bridge, or any gallery provided for similar purposes or for maintenance and repair);

e. an external wall; and

f. a compartment wall (including a wall common to two or more buildings). (However, see the guidance to B3, paragraph 8.4, for exclusions from the provisions for elements of structure).

Emergency lighting Lighting provided for use when the supply to the normal lighting fails.

Escape lighting That part of the emergency lighting which is provided to ensure that the escape route is illuminated at all material times.

Escape route Route forming that part of the means of escape from any point in a building to a final exit.

European Technical Approval A favourable technical assessment of the fitness for use of a construction product for an intended use, issued for the purposes of the Construction Products Directive by a body authorised by a member State to issue European technical approvals for those purposes and notified by that member State to the European Commission.

European Technical Approvals Issuing body A body notified under article 10 of the Construction Products Directive. The details of these institutions are published in the "C" series of the Official Journal of the European Communities. (At the present time the listing for the United Kingdom is the British Board of Agrément and WIMLAS Ltd. An up to date listing can be found on the Building Regulations pages of the DETR Website at http://www.detr.gov.uk/)

Evacuation lift A lift that may be used for the evacuation of disabled people in a fire.

External wall (or side of a building) includes a part of a roof pitched at an angle of more than 70° to the horizontal, if that part of the roof adjoins a space within the building to which persons have access (but not access only for repair or maintenance).

Final exit The termination of an escape route from a building giving direct access to a street, passageway, walkway or open space, and sited to ensure the rapid dispersal of persons from the vicinity of a building so that they are no longer in danger from fire and/or smoke.

> **Note:** Windows are not acceptable as final exits.

Fire door A door or shutter, provided for the passage of persons, air or objects, which together with its frame and furniture as installed in a building, is intended (when closed) to resist the passage of fire and/or gaseous products of combustion, and is capable of meeting specified performance criteria to those ends. (It may have one or more leaves, and the term includes a cover or other form of protection to an opening in a fire-resisting wall or floor, or in a structure surrounding a protected shaft).

Firefighting lift A lift designed to have additional protection, with controls that enable it to be used under the direct control of the fire service in fighting a fire. (See Sections 16-19).

Firefighting lobby A protected lobby providing access from a firefighting stair to the accommodation area and to any associated firefighting lift.

Firefighting shaft A protected enclosure containing a firefighting stair, firefighting lobbies and, if provided, a firefighting lift, together with its machine room.

Firefighting stair A protected stairway communicating with the accommodation area only through a firefighting lobby.

Fire-resisting (fire resistance) The ability of a component or construction of a building to satisfy for a stated period of time, some or all of the appropriate criteria specified in the relevant Part of BS 476.

Fire separating element A compartment wall, compartment floor, cavity barrier and construction enclosing a protected escape route and/or a place of special fire hazard.

Fire stop A seal provided to close an imperfection of fit or design tolerance between elements or components, to restrict the passage of fire and smoke.

* Flat A separate and self-contained premises constructed or adapted for use for residential purposes and forming part of a building from some other part of which it is divided horizontally.

Gallery A floor which is less than one-half of the area of the space into which it projects.

Habitable room A room used, or intended to be used, for dwelling purposes (including for the purposes of Part B, a kitchen, but not a bathroom).

Height (of a building or storey for the purposes of Part B) Height of a building is measured as shown in Appendix C, Diagram C3, and height of the floor of the a top storey above ground is measured as shown in Appendix C, Diagram C5.

Inner room Room from which escape is possible only by passing through another room (the access room).

Maisonette means a 'Flat' on more than one level.

Material of limited combustibility A material performance specification that includes non-combustible materials, and for which the relevant test criteria are set out in Appendix A, paragraph 9.

Means of escape Structural means whereby [in the event of fire] a safe route or routes is or are provided for persons to travel from any point in a building to a place of safety.

Measurement Area, cubic capacity, height of a building and number of storeys, see Appendix C, Diagrams C1 to C5; occupant capacity, travel distance, and width of a doorway, escape route and a stair, see paragraph B1.xxv.

Non-combustible material The highest level of reaction to fire performance. The relevant test criteria are set out in Appendix A, paragraph 8.

Notional boundary A boundary presumed to exist between buildings on the same site (see Section 14, Diagram 42). The concept is applied only to buildings in the residential and the assembly and recreation purpose groups.

Occupancy type A purpose group identified in Appendix D.

Open spatial planning The internal arrangement of a building in which more than one storey or level is contained in one undivided volume, eg split-level floors. For the purposes of this document there is a distinction between open spatial planning and an atrium space.

Perimeter (of building) The maximum aggregate plan perimeter, found by vertical projection onto a horizontal plane (see Section 17, Diagram 48).

Pipe (for the purposes of Section 11) – includes pipe fittings and accessories; and excludes a flue pipe and a pipe used for ventilating purposes (other than a ventilating pipe for an above around drainage system).

Places of special fire hazard Oil-filled transformer and switch gear rooms, boiler rooms, storage space for fuel or other highly flammable substances, and rooms housing a fixed internal combustion engine and (additionally in schools) laboratories, technology rooms with open heat sources, kitchens and stores for PE mats or chemicals.

Platform floor (access or raised floor) A floor supported by a structural floor, but with an intervening concealed space which is intended to house services.

Protected circuit An electrical circuit protected against fire.

Protected corridor/lobby A corridor or lobby which is adequately protected from fire in adjoining accommodation by fire-resisting construction.

Protected entrance hall/landing A circulation area consisting of a hall or space in a dwelling, enclosed with fire-resisting construction (other than any part which is an external wall of a building).

Protected shaft A shaft which enables persons, air or objects to pass from one compartment to another, and which is enclosed with fire-resisting construction.

Protected stairway A stair discharging through a final exit to a place of safety (including any exit passageway between the foot of the stair and the final exit) that is adequately enclosed with fire-resisting construction.

Purpose group A classification of a building according to the purpose to which it is intended to be put. See Appendix D, Table D1.

Relevant boundary The boundary which the side of the building faces, (and/or coincides with) and which is parallel, or at an angle of not more than 80°, to the side of the building (see Section 14 Diagram 41). A notional boundary can be a relevant boundary.

Rooflight A dome light, lantern light, skylight, ridge light, glazed barrel vault or other element intended to admit daylight through a roof.

Room (for the purposes of B2) An enclosed space within a building that is not used solely as a circulation space. (The term includes not only conventional rooms, but also cupboards that are not fittings, and large spaces such as warehouses, and auditoria. The term does not include voids such as ducts, ceiling voids and roof spaces).

Separated part (of a building) A form of compartmentation in which a part of a building is separated from another part of the same building by a compartment wall. The wall runs the full height of the part, and is in one vertical plane. (See paragraph 9.24, and Appendix C, Diagram C4).

Single storey building A building consisting of a ground storey only. (A separated part which consists of a ground storey only, with a roof to which access is only provided for repair or maintenance, may be treated as a single storey building). Basements are not included in counting the number of storeys in a building (see Appendix C).

Site (of a building) is the land occupied by the building, up to the boundaries with land in other ownership.

Smoke alarm A device containing within one housing all the components, except possibly the energy source, necessary for detecting smoke and giving an audible alarm.

Storey includes:

a. any gallery in an assembly building (Purpose Group 5); and

b. any gallery in any other type of building if its area is more than half that of the space into which it projects; and

c. a roof, unless it is accessible only for maintenance and repair.

Storey exit A final exit, or a doorway giving direct access into a protected stairway, firefighting lobby, or external escape route.

  Note: A door in a compartment wall in an institutional building is considered as a storey exit for the purposes of B1 if the building is planned for progressive horizontal evacuation, see paragraph 4.30.

Suspended ceiling (fire-protecting) A ceiling suspended below a floor, which contributes to the fire resistance of the floor. Appendix A, Table A3, classifies different types of suspended ceiling.

Technical specification A standard or a European Technical Approval Guide. It is the document against which compliance can be shown in the case of a standard and against which an assessment is made to deliver the European technical approval.

Thermoplastic material See Appendix A, paragraph 17.

Travel distance (unless otherwise specified, eg as in the case of flats) The actual distance to be travelled by a person from any point within the floor area to the nearest storey exit, having regard to the layout of walls, partitions and fittings.

Unprotected area In relation to a side or external wall of a building means:

a. window, door or other opening; and

  Note: Windows that are not openable and are designed and glazed to provide the necessary level of fire resistance, and recessed car parking areas shown in Diagram E1, need not be regarded as an unprotected area.

b. any part of the external wall which has less than the relevant fire resistance set out in Section 13; and

c. any part of the external wall which has combustible material more than 1mm thick attached or applied to its external face, whether for cladding or any other purpose. (Combustible material in this context is any material which does not have a Class 0 rating).

Diagram E1 **Recessed car parking areas**

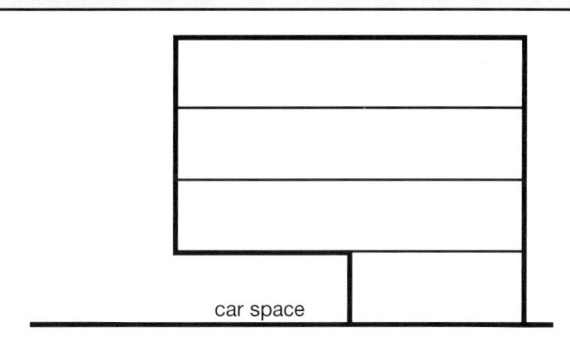

Note: The parking area should be:

a   open fronted

b   separated from the remainder of the building by a compartment wall(s) and floor(s) having not less than the period of fire resistance specified in Table A2 in Appendix A.

Appendix F

FIRE BEHAVIOUR OF INSULATING CORE PANELS USED FOR INTERNAL STRUCTURES

Introduction

1 Insulating core panel systems are used for external cladding as well as for internal structures. However, whilst both types of panel system have unique fire behaviour characteristics, it is those used for internal structures that can present particular problems with regard to fire spread.

The most common use of insulating core panels, when used for internal structures, is to provide an enclosure in which a chilled or sub zero environment can be generated for the production, preservation, storage and distribution of perishable foodstuffs. However this type of construction is also used in many other applications, particularly where the maintenance of a hygienic environment is essential.

These panels typically consist of an inner core sandwiched between, and bonded to, a membrane such as facing sheets of galvanised steel, often bonded with a PVC facing for hygiene purposes. The panels are then formed into a structure by jointing systems, usually designed to provide an insulating and hygienic performance. The panel structure can be free standing, but is usually attached to the building structure by lightweight fixings and hangers.

The most common forms of insulation in present use are:

• expanded polystyrene,

• extruded polystyrene,

• polyurethane,

• mineral fibre.

However panels with the following core materials are also in use:

• polyisocyanurate,

• modified phenolic.

Fire behaviour of the core materials and fixing systems

2 The degradation of polymeric materials can be expected when exposed to radiated/conducted heat from a fire, with the resulting production of large quantities of smoke.

It is recognised that the potential for problems in fires involving mineral fibre cores is less than those for polymeric core materials.

In addition, irrespective of the type of core material, the panel, when exposed to the high temperatures of a developed fire, will tend to delaminate between the facing and core material, due to a combination of expansion of the membrane and softening of the bond line.

Therefore once it is involved, either directly or indirectly in a fire, the panel will have lost most of its structural integrity. The stability of the system will then depend on the residual structural strength of the non-exposed facing, the joint between panels and the fixing system.

Most jointing or fixing systems for these systems have an extremely limited structural integrity performance in fire conditions. If the fire starts to heat up the support fixings or structure to which they are attached, then there is a real chance of total collapse of the panel system.

The insulating nature of these panels, together with their sealed joints, means that fire can spread behind the panels, hidden from the occupants of occupied rooms/spaces.

This can prove to be a particular problem to fire fighters as, due to the insulating properties of the cores, it may not be possible to track the spread of fire, even using infra red detection equipment. This difficulty, together with that of controlling the fire spread within and behind the panels, is likely to have a detrimental effect on the performance of the fixing systems, potentially leading to their complete and unexpected collapse, together with any associated equipment.

Fire fighting

3 When compared with other types of construction techniques, these panel systems therefore provide a unique combination of problems for fire fighters, including:

• hidden fire spread within the panels,

• production of large quantities of black toxic smoke, and

• rapid fire spread leading to flashover.

These three characteristics are common to both polyurethane and polystyrene cored panels, although the rate of fire spread in polyurethane cores is significantly less than that of polystyrene cores, especially when any external heat source is removed.

In addition, irrespective of the type of panel core, all systems are susceptible to:

• delamination of the steel facing,

• collapse of the system,

• hidden fire spread behind the system.

Design recommendations

4 To identify the appropriate solution, a risk assessment approach should be adopted. This would involve identifying the potential fire risk within the enclosures formed by the panel systems and then adopting one or more of the following at the design stage:

- removing the risk,
- separating the risk from the panels by an appropriate distance,
- providing a fire suppression system for the risk,
- providing a fire suppression system for the enclosure,
- providing fire-resisting panels,
- specifying appropriate materials/fixing and jointing systems.

In summary the performance of the building structure, including the insulating envelope, the superstructure, the substructure etc, must be considered in relation to their performance in the event of a fire.

Specifying panel core materials

5 Where at all possible the specification of panels with core materials appropriate to the application will help ensure an acceptable level of performance for panel systems, when involved in fire conditions.

The following are examples in the provision of core materials which may be appropriate to the application concerned.

Mineral fibre cores:

- cooking areas,
- hot areas,
- bakeries,
- fire breaks in combustible panels,
- fire stop panels,
- general fire protection.

All cores:

- chill stores,
- cold stores,
- blast freezers,
- food factories,
- clean rooms.

Note: Core materials may be used in other circumstances where a risk assessment has been made and other appropriate fire precautions have been put in place.

Specifying materials/fixing and jointing systems

6 The following are methods by which the stability of panel systems may be improved in the event of a fire, although they may not all be appropriate in every case.

In addition the details of construction of the insulating envelope should, particularly in relation to combustible insulant cores, prevent the core materials from becoming exposed to the fire and contributing to the fire load.

a. Insulating envelopes, support systems, and supporting structure should be designed to allow the envelope to remain structurally stable by alternative means such as catenary action following failure of the bond line between insulant core and facing materials. This will typically require positive attachment of the lower faces of the insulant panels to supports.

b. The building superstructure, together with any elements providing support to the insulating envelope, should be protected to prevent early collapse of the structure or the envelope.

   Note: Irrespective of the type of panel provided, it will remain necessary to ensure that the supplementary support method supporting the panels remains stable for an appropriate time period under fire conditions. It is not practical to fire protect light gauge steel members such as purlins and sheeting rails which provide stability to building superstructures and these may be compromised at an early stage of a fire. Supplementary fire protected heavier gauge steelwork members could be provided at wider intervals than purlins to provide restraint in the event of a fire.

c. In designated high risk areas, consideration should be given to incorporating non-combustible insulant cored panels into wall and ceiling construction at intervals, or incorporating strips of non-combustible material into specified wall and ceiling panels, in order to provide a barrier to fire propagation through the insulant.

d. Correct detailing of the insulating envelope should ensure that the combustible insulant is fully encapsulated by non-combustible facing materials which remain in place during a fire.

e. The panels should incorporate pre-finished and sealed areas for penetration of services.

General

7 Panels or panel systems should not be used to support machinery or other permanent loads.

Any cavity created by the arrangement of panels, their supporting structure or other building elements should be provided with suitable cavity barriers.

8 Examples of possible solutions and general guidance on insulating core panels construction can be found in *Design, construction, specification and fire management of insulated envelopes for temperature controlled environments* published by the International Association of Cold Storage Contractors (European Division).

Of particular relevance is Chapter 8 of the document which gives guidance on the design, construction and management of insulated structures. Whilst the document is primarily intended for use in relation to cold storage environments, the guidance, particularly in Chapter 8, is considered to be appropriate for most insulating core panel applications.

Appendix G

STANDARDS REFERRED TO

General Introduction

BS 5588: *Fire precautions in the design, construction and use of buildings:*
Part 7: 1997 *Code of practice for the incorporation of atria in buildings.*
Part 10: 1991 *Code of practice for shopping complexes.*

DD 240: 1997 Fire safety engineering in buildings.

Approved Document B1

BS EN 54-11 Fire detection and fire alarm systems – Part 11: Manual call points.

BS 5266: *Emergency lighting:*
Part 1: 1988 *Code of practice for the emergency lighting of premises other than cinemas and certain other specified premises used for entertainment.*

BS 5306: *Fire extinguishing installations and equipment on premises:*
Part 2: 1990 *Specification for sprinkler systems.*

BS 5395: *Stairs, ladders and walkways:*
Part 2: 1984 *Code of practice for the design of helical and spiral stairs*
Amendment slip
1: AMD 6076

BS 5446: *Components of automatic fire alarm systems for residential premises:*
Part 1: 1990 *Specification for self-contained smoke alarms and point-type smoke detectors.*

BS 5449: *Fire safety signs, notices and graphic symbols:*
Part 1: 1990 *Specification for fire safety signs.*

BS 5588: *Fire precautions in the design, construction and use of buildings:*
Part 0: 1996 *Guide to fire safety codes of practice for particular premises/applications.*
Part 1: 1990 *Code of practice for residential buildings*
Part 4: 1998 *Code of practice for smoke control using pressure differentials.*
Part 5: 1991 *Code of practice for firefighting stairs and lifts.*
Part 6: 1991 *Code of practice for places of assembly.*
Part 7: 1997 *Code of practice for the incorporation of atria in buildings.*
Part 8: 1999 *Code of practice for means of escape for disabled people.*
Part 9: 1989 *Code of practice for ventilation and air conditioning ductwork.*
Part 10: 1991 *Code of practice for shopping complexes.*
Part 11: 1997 *Code of practice for shops, offices, industrial, storage and other similar buildings.*

BS 5720: 1979 *Code of practice for mechanical ventilation and air conditioning in buildings.*

BS 5839: *Fire detection and alarm systems for buildings:*
Part 1: 1988 *Code of practice for system design, installation and servicing.*
Part 2: 1983 *Specification for manual call points*
Part 6: 1995 *Code of practice for the design and installation of fire detection and alarm systems in dwellings.*
Part 8: 1998 *Code of practice for the design, installation and servicing of voice alarm systems.*

BS 5906: 1980 *Code of practice for storage and on-site treatment of solid waste from buildings.*

BS 6387: 1994 *Specification for performance requirements for cables required to maintain circuit integrity under fire conditions.*

CP 1007: 1955 *Maintained lighting for cinemas.*

Approved Document B2

BS 476: *Fire tests on building materials and structures.*
Part 4: 1970 (1984) *Non-combustibility test for materials.*
Amendment slips
1: AMD 2483,
2: AMD 4390.
Part 6: 1981 *Method of test for fire propagation for products.*
Part 6: 1989 *Method of test for fire propagation for products.*
Part 7: 1971 *Surface spread of flame test for materials.*
Part 7: 1987 *Method for classification of the surface spread of flame of products.*
Amendment slip
1: AMD 6249
Part 7: 1997 *Method of test to determine the classification of the surface spread of flame of products.*
Part 11: 1982 *Method for assessing the heat emission from building materials.*

BS 6661: 1986 *Guide for design, construction and maintenance of single-skin air supported structures.*

BS 7157: 1989 *Method of test for ignitability of fabrics used in the construction of large tented structures.*

Approved Document B3

BS 4514: 1983 *Specification for unplasticized PVC soil and ventilating pipes, fittings and accessories.*
Amendment slips
1: AMD 4517
2: AMD 5584

BS 5255: 1989 *Specification for thermoplastics waste pipe and fittings.*

BS 5306: *Fire extinguishing installations and equipment on premises:*
Part 2: 1990 *Specification for sprinkler systems.*

BS 5588: *Fire precautions in the design, construction and use of buildings:*
Part 5: 1991 *Code of practice for firefighting stairs and lifts.*
Part 7: 1997 *Code of practice for the incorporation of atria in buildings.*
Part 9: 1989 *Code of practice for ventilation and air conditioning ductwork.*
Part 10: 1991 *Code of practice for shopping complexes.*

BS 5839: *Fire detection and alarm systems for buildings:*
Part 1: 1988 *Code of practice for system design, installation and servicing.*

BS 7346: *Components for smoke and heat control systems:*
Part 2: 1990 *Specification for powered smoke and heat exhaust ventilators*

BS 8313: 1989 *Code of practice for accommodation of building services in ducts*

Approved Document B4

BS 476: *Fire tests on building materials and structures:*
Part 3: 1958 *External fire exposure roof tests.*
Part 6: 1981 *Method of test for fire propagation for products.*
Part 6: 1989 *Method of test for fire propagation for products.*
Part 7: 1971 *Surface spread of flame tests for materials.*
Part 7: 1987 *Method for classification of the surface spread of flame of products.*
Amendment slip
1: AMD 6249.
Part 7: 1997 *Method of test to determine the classification of the surface spread of flame of products.*

BS 5306: *Fire extinguishing installations and equipment on premises:*
Part 2: 1990 *Specification for sprinkler systems.*

BS 5588: *Fire precautions in the design, construction and use of buildings*
Part 5: 1991 *Code of practice for firefighting stairs and lifts.*
Part 7: 1997 *Code of practice for the incorporation of atria in buildings.*

Approved Document B5

BS 5306: *Fire extinguishing installations and equipment on premises:*
Part 1: 1976 (1988) *Hydrant systems, hose reels and foam inlets.*
Amendment slips
1: AMD 4649,
2: AMD 5766.
Part 2: 1990 *Specification for sprinkler systems.*

BS 5588: *Fire precautions in the design, construction and use of buildings:*
Part 5: 1991 *Code of practice for firefighting stairs and lifts.*
Part 10: 1991 *Code of practice for shopping complexes.*

BS 5839: *Fire detection and alarm systems for buildings:*
Part 1: 1988 *Code of practice for system design, installation and servicing.*

BS 7346: *Components for smoke and heat control systems:*
Part 2: 1990 *Specification for powered smoke and heat exhaust ventilators*

Appendix A

BS 476: *Fire tests on building materials and structures:*
Part 3: 1958 *External fire exposure roof tests.*
Part 4: 1970 (1984) *Non-combustibility test for materials.*
Amendment slips
1: AMD 2483,
2: AMD 4390.
Part 6: 1981 *Method of test for fire propagation for products.*
Part 6: 1989 *Method of test for fire propagation for products.*
Part 7: 1971 *Surface spread of flame tests for materials.*
Part 7: 1987 *Method for classification of the surface spread of flame of products.*
Amendment slip
1: AMD 6249.
Part 7: 1997 *Method of test to determine the classification of the surface spread of flame of products.*
Part 8: 1972 *Test methods and criteria for the fire resistance of elements of building construction.*
Amendment slips
1: AMD 1873,
2: AMD 3816,
3: AMD 4822.
Part 11: 1982 *Method for assessing the heat emission from building products.*
Part 20: 1987 *Method for determination of the fire resistance of elements of construction (general principles).*
Amendment slip
1: AMD 6487.
Part 21: 1987 *Methods for determination of the fire resistance of loadbearing elements of construction.*
Part 22: 1987 *Methods for determination of the fire resistance of non-loadbearing elements of construction.*
Part 23: 1987 *Methods for determination of the contribution of components to the fire resistance of a structure.*
Part 24: 1987 *Method for determination of the fire resistance of ventilation ducts.*

BS 2782: 1970 *Methods of testing plastics:*
Part 5: Miscellaneous methods: Method 508A.
Rate of burning (Laboratory method).

BS 2782: *Methods of testing plastics.*
Part 1: *Thermal properties:*
Methods 120A to 120E: 1990 *Determination of the Vicat softening temperature of thermoplastics.*

BS 5306: *Fire extinguishing installations and equipment on premises:*
Part 2: 1990 *Specification for sprinkler systems.*

BS 5438: 1989 *Methods of test for flammability of textile fabrics when subjected to a small igniting flame applied to the face or bottom edge of vertically oriented specimens,* Test 2.

BS 5588: *Fire precautions in the design, construction and use of buildings*
Part 7: 1997 *Code of practice for the incorporation of atria in buildings.*
Part 8: 1999 *Code of practice for means of escape for disabled people.*

BS 5867: *Specification for fabrics for curtains and drapes:*
Part 2: 1980 *Flammability requirements.*
Amendment slip
1: AMD 4319.

BS 6073: *Precast concrete masonry units:*
Part 1: 1981 *Specification for precast concrete masonry units.*
Amendment slips
1: AMD 3944,
2: AMD 4462.

BS 6336: 1998 *Guide to development and presentation of fire tests and their use in hazard assessment.*

PD 6520: 1988 *Guide to fire test methods for building materials and elements of construction.*

European test methods and classifications (Reaction to fire)

BS EN ISO 1182:2002, *Reaction to fire tests for building products - Non-combustibility test.*

BS EN ISO 1716:2002, *Reaction to fire tests for building products - Determination of the gross calorific value.*

BS EN 13823:2002, *Reaction to fire tests for building products - Building products excluding floorings exposed to the thermal attack by a single burning item.*

BS EN ISO 11925-2:2002, *Reaction to fire tests for building Products, Part 2-Ignitability when subjected to direct impingement of a flame.*

BS EN 13238:2001, *Reaction to fire tests for building products-conditioning procedures and general rules for selection of substrates.*

BS EN 13501-1:2002, *Fire classification of construction products and building elements, Part 1- Classification using data from reaction to fire tests.*

European test methods and classifications (Fire resistance)

BS EN 1363-1:1999, *Fire resistance tests, Part 1- General requirements.*

BS EN 1363-2:1999, *Fire resistance tests, Part 2- Alternative and additional procedures.*

DD ENV 1363- 3:1999, *Fire resistance tests, Part 3-Verification of furnace performance.*

BS EN 1364-1:1999, *Fire resistance tests for non-loadbearing elements, Part 1-Walls.*

BS EN 1364-2:1999, *Fire resistance tests for non-loadbearing elements, Part 2-Ceilings.*

BS EN 1364- 3:xxxx, *Fire resistance tests for non-loadbearing elements, Part 3-Curtain walls-full configuration.*

BS EN 1364-4:xxxx, *Fire resistance tests for non-loadbearing elements, Part 4-Curtain walls-part configuration.*

BS EN 1364-5:xxxx, *Fire resistance tests for non-loadbearing elements, Part 5-Semi-natural fire test for facades and curtain walls.*

BS EN 1364- 6:xxxx, *Fire resistance tests for non-loadbearing elements, Part 6-External wall systems.*

BS EN 1365-1:1999, *Fire resistance tests for loadbearing elements, Part 1-Walls.*

BS EN 1365-2:1999, *Fire resistance tests for loadbearing elements, Part 2-Floors and roofs.*

BS EN 1365- 3:1999, *Fire resistance tests for loadbearing elements, Part 3-Beams.*

BS EN 1365-4:xxxx, *Fire resistance tests for loadbearing elements, Part 4-Columns.*

BS EN 1365-5:xxxx, *Fire resistance tests for loadbearing elements, Part 5-Balconies.*

BS EN 1365- 6:xxxx, *Fire resistance tests for loadbearing elements, Part 6-Stairs and walkways.*

BS EN 1366-1:1999, *Fire resistance tests for service installations, Part 1-Ducts.*

BS EN 1366-2:1999, *Fire resistance tests for service installations, Part 2-Fire dampers.*

BS EN 1366- 3:xxxx, *Fire resistance tests for service installations, Part 3-Penetration seals.*

BS EN 1366-4:xxxx, *Fire resistance tests for service installations, Part 4-Linear joint seals.*

BS EN 1366-5:xxxx, *Fire resistance tests for service installations, Part 5-Service ducts and shafts.*

BS EN 1366- 6:xxxx, *Fire resistance tests for service installations, Part 6-Raised floors.*

BS EN 1366-7:xxxx, *Fire resistance tests for service installations, Part 7-Closures for conveyors and trackbound transportation systems.*

BS EN 1366-8:xxxx, *Fire resistance tests for service installations, Part 8-Smoke extraction ducts.*

BS EN 1366-9:xxxx, *Fire resistance tests for service installations, Part 9-Single compartment smoke extraction ducts.*

BS EN 1366-10:xxxx, *Fire resistance tests for service installations, Part 10-Smoke control dampers.*

BS EN 13501-2:xxxx, *Fire classification of construction products and building elements, Part 2- Classification using data from fire resistance tests (excluding products for use in ventilation systems).*

BS EN 13501- 3:xxxx, *Fire classification of construction products and building elements, Part 3-Classification using data from fire resistance tests on components of normal building service installations (other than smoke control systems).*

BS EN 13501-4:xxxx, *Fire classification of construction products and building elements, Part 4-Classification using data from fire resistance tests on smoke control systems.*

DD ENV 13381-1:xxxx, *Test methods for determining the contribution to the fire resistance of structural members, Part 1-Horizontal protective membranes.*

DD ENV 13381-2:2002, *Test methods for determining the contribution to the fire resistance of structural members, Part 2-Vertical protective membranes.*

DD ENV 13381- 3:2002, *Test methods for determining the contribution to the fire resistance of structural members, Part 3- Applied protection to concrete members.*

DD ENV 13381-4:2002, *Test methods for determining the contribution to the fire resistance of structural members, Part 4- Applied protection to steel members.*

DD ENV 13381-5:2002, *Test methods for determining the contribution to the fire resistance of structural members, Part 5- Applied protection to concrete/ profiled sheet steel composite members.*

DD ENV 13381- 6:2002, *Test methods for determining the contribution to the fire resistance of structural members, Part 6-Applied protection to concrete filled hollow steel columns.*

DD ENV 13381-7:2002, *Test methods for determining the contribution to the fire resistance of structural members, Part 7-Applied protection to timber members.*

European test methods and classifications (External fire exposure of roofs)

DD ENV 1187:2002, *Test methods for external fire exposure to roofs.*

Appendix B

BS 476: *Fire tests on building materials and structures:*
Part 8: 1972 *Test methods and criteria for the fire resistance of elements of building construction.*
Amendment slips
1: AMD 1873,
2: AMD 3816,
3: AMD 4822.
Part 22: 1987 *Methods for determination of the fire resistance of non-loadbearing elements of construction.*
Part 31: *Methods for measuring smoke penetration through doorsets and shutter assemblies:*

Section 31.1: 1983 *Measurement under ambient temperature conditions.*
Amendment slip
1. AMD 8366.

BS 5499: *Fire safety signs, notices and graphic symbols:*
Part 1: 1990 *Specification for fire safety signs.*

BS 5588: *Fire precautions in the design, construction and use of buildings:*
Part 4: 1998 *Code of practice for smoke control using pressure differentials.*

BS 8214: 1990 *Code of practice for fire door assemblies with non-metallic leaves.*

BS EN 1634-1:2000, *Fire resistance tests for door and shutter assemblies, Part 1-Fire doors and shutters.*

BS EN 1634-2:xxxx, *Fire resistance tests for door and shutter assemblies, Part 2-Fire door hardware.*

BS EN 1634- 3:xxxx, *Fire resistance tests for door and shutter assemblies, Part 3-Smoke control doors.*

OTHER PUBLICATIONS REFERRED TO

Use of guidance

Construction Products Directive (CPD). The Council Directive reference 89/106/EEC dated 21 December 1988 and published in the Official Journal of the European Communities No L40/12 dated 11.2.89. The CE Marking Directive (93/68/EEC) amends the CPD.

Construction fire safety, Construction Information Sheet No 51.

Fire safety in construction work, HSG 168 (ISBN 0-7176-1332-1).

Workplace health, safety and welfare. The Workplace (Health, Safety and Welfare) Regulations 1992, Approved Code of Practice and Guidance; The Health and Safety Commission, L24; HMSO 1992; ISBN 0-11-886333-9.

General Introduction

Crown Fire Standards. Property Advisers to the Civil Estate (PACE). (Available from Corporate Documents Services, Savile House, Trinity Arcade, Leeds, LS1 6QW)

Firecode. HTM 81. Fire precautions in new hospitals. (NHS Estates) HMSO, 1996.

LPC Design guide for the fire protection of buildings. Loss Prevention Council, 1996.

Approved Document B1

The Building Regulations 1991 (DETR/Welsh Office) HMSO
Approved Document K, Protection from falling, collision and impact;
Approved Document M, Access and facilities for disabled people;
Approved Document N, Glazing – safety in relation to impact, opening and cleaning

Building Regulation and Fire Safety – Procedural Guidance. (DOE/Home Office/Welsh Office 1992)

Design methodologies for smoke and heat exhaust ventilation. BR 368, BRE 1999. (Revision of Design principles for smoke ventilation in enclosed shopping centres. BR 186, BRE, 1990.)

DOE Circular 12/92. Houses in multiple occupation. Guidance to local housing authorities on standards of fitness under section 352 of the Housing Act 1985. HMSO, 1992.

DOE Circular 12/93. Houses in multiple occupation. Guidance to local housing authorities on managing the stock in their area. HMSO, 1993.

Draft guide to fire precautions in existing residential care premises. Home Office/Scottish Home and Health Department, 1983.

Firecode. HTM 81. Fire precautions in new hospitals. (NHS Estates) HMSO, 1996.

Firecode. HTM 88. Guide to fire precautions in

NHS housing in the community for mentally handicapped (or mentally ill) people. (DHSS) HMSO, 1986.

Fire Precautions Act 1971. Guide to fire precautions in existing places of work that require a fire certificate. Factories, offices, shops and railway premises. (Home Office/Scottish Office) HMSO, 1993.

Fire Precautions (Workplace) Regulations 1997, (SI 1997 No 1840) as amended by the Fire Precautions (Workplace) (Amendment) Regulations 1999.

Gas Safety (Installation and Use) Regulations 1998, SI 1998 No 2451.

Guide to fire precautions in existing places of entertainment and like premises. (Home Office/ Scottish Home and Health Department) HMSO, 1990.

Guide to Safety at Sports Grounds. (Department of National Heritage/Scottish Office) HMSO, 1997.

Pipelines Safety Regulations 1996, SI 1996 No 825

Safety signs and signals. The Health and Safety (Safety signs and signals) Regulations 1996. Guidance on Regulations. (HSE, L64). HSE Books,1996.

Welsh Office Circular 25/92. Local Government and Housing Act 1989. Houses in multiple occupation: standards of fitness. Welsh Office, 1992.

Welsh Office Circular 55/93. Houses in multiple occupation. Guidance on management strategies. Welsh Office, 1993.

Approved Document B2

The Building Regulations 1991. Approved Document N, Glazing – safety in relation to impact, opening and cleaning (DETR/Welsh Office) HMSO.

Fire safety of PTFE-based materials used in buildings. BR 274, BRE 1994.

Approved Document B3

The Building Regulations 1991(DETR/Welsh Office) HMSO.
Approved Document F, Ventilation
Approved Document J, Combustion appliances and fuel storage

Design methodologies for smoke and heat exhaust ventilation. BR 368, BRE 1999. (Revision of Design principles for smoke ventilation in enclosed shopping centres. BR 186, BRE, 1990.)

Gas Safety (Installation and Use) Regulations 1998, SI 1998 No 2451.

Pipelines Safety Regulations 1996, SI 1996 No 825

Approved Document B4

The Building Regulations 1991(DETR/Welsh Office) HMSO. Approved Document J, Combustion appliances and fuel storage

Assessing the fire performance of external cladding systems: a test method, BRE Fire Note 9 (BRE, 1999).

External fire spread: Building separation and boundary distances. BR 187, BRE, 1991.

Fire performance of external thermal insulation for walls of multi-storey buildings. BR 135, BRE 1988.

Fire and steel construction: The behaviour of steel portal frames in boundary conditions, 1990. 2nd edition (available from the Steel Construction Institute, Silwood Park, Ascot, Berks, SL5 7QN).

Heat radiation from fires and building separation. Fire Research Technical Paper 5. HMSO, 1963.

Thatched buildings. New properties and extensions [the "Dorset Model"] (can be found on www.dorset-technical-committee.org.uk; email: info@dorset-technical-committee.org.uk)

Appendix A

Fire protection for structural steel in buildings (second edition – revised). ASFP/SCI/FTSG, 1992 (available from the ASFP, Association House, 99 West Street, Farnham, Surrey GU9 7EN and the Steel Construction Institute, Silwood Park, Ascot, Berks SL5 7QN).

Guidelines for the construction of fire-resisting structural elements. BR 128, BRE, 1988.

Increasing the fire resistance of existing timber floors. BRE Digest 208, 1988.

Appendix B

Hardware for timber and escape doors published by the Builders Hardware Industry Federation in November 2000 is available from the BHIF, 42 Heath Street, Tamworth, Staffordshire, B79 7JH.

Code of practice for fire-resisting metal doorsets published by the DSMA (Door and Shutter Manufacturers' Association) in 1999 is available from DSMA, 42 Heath Street, Tamworth, Staffs B79 7JH.

Timber Fire Resisting Doorsets: maintaining performance under the new European test standard published by TRADA is available from TRADA Technology Limited, Stocking Lane, Hughenden Valley, High Wycombe, Bucks, HP14 4ND.

Appendix F

Design, construction, specification and fire management of insulated envelopes for temperature controlled environments. The International Association of Cold Storage Contractors (European Division), 1999. (Available from the IACSC, Downmill Road, Bracknell, Berks RG12 1GH.)

Index

A

Above ground drainage
See Drainage
Access
See Fire service facilities
Access corridors
See Corridor access
Access for fire service
See Fire service facilities
Access panels
Fire protecting suspended ceilings
table A3
Access rooms
Definition appendix E
Escape from flats and
maisonettes 3.7
Escape in single family houses 2.4
Horizontal escape from buildings
other than dwellings 4.9,
diagram 17
Accommodation stairs 5.23
Definition appendix E
**Active measures for fire
extinguishment** 0.14
See also Sprinkler systems
Aggregate notional area
Space separation 14.15
Air changes
See Ventilation
**Air circulation systems for
heating, energy conservation
or condensation control**
Flats and maisonettes with a floor
more than 4.5m above ground
level 3.16
Houses with a floor more than
4.5m above ground level 2.15
Air conditioning 6.49
Ducts and integrity of
compartments 11.10
See also Ventilation
Air supported structures 7.8
Aisles
Fixed seating 4.18, 6.30
Alterations
Material alteration 0.19
Alternative approaches B1.xviii
Alternative escape routes B1.xi, B1.xiii
Definition appendix E
Flats and maisonettes 3.18
Horizontal escape from buildings
other than dwellings 4.3, 4.5, 4.8,
4.23, diagram 16
Houses with floor more than
4.5m above ground 2.12, 2.13
Provision of cavity barriers table 13
Alternative exits
Definition appendix E
Fire doors table B1
Flats and maisonettes
Balconies 3.9
Floor more than 4.5m above
ground 3.11-3.15, diagrams 9-10
Progressive horizontal
evacuation 4.30
Amusement arcades
Floor space factors table 1
Ancillary accommodation
Escape routes for flats or
maisonettes 3.27, 3.43-3.44
Ancillary use
Purpose groups appendix D(3)-(5)
Anticipated fire severity 0.13
Appliance ventilation ducts
Compartments 9.35, 11.11
Definition appendix E

Approved documents
page 5
Architectural liaison officers
Source of advice on security B1.xvii
Architraves
Definition of wall and ceiling 7.2-7.3
Area
Methods of measurement diagram C2
Art galleries
Floor space factors table 1
Artificial lighting
See Lighting
**Assembly and recreation
purpose group**
Automatic fire detection and alarm
systems 1.31
Compartmentation 9.20
Counting number of storeys
diagram C4
Escape lighting table 9
Escape route design 4.4
   Spacing of fixed seating 4.18
Travel distance limitations table 3
External escape stairs 5.33
External walls 13.2, diagram 40
Fire doors table B1
Floor space factors table 1
Junction of compartment wall
with roof 9.30
Maximum dimensions of building
or compartment table 12
Means of escape B1.xxii
Mechanical ventilation 6.47
Minimum width of escape
stairs table 6
Panic bolts 6.12
Provision of cavity barriers table 13
Purpose groups table D1
Simultaneous evacuation 5.15
Space separation 14.2
   Acceptable unprotected
   areas 14.16
   Notional boundaries 14.6
   Permitted unprotected areas in
   small buildings or compartments
   table 16
Atria
Automatic fire detection and alarm
systems 1.31
Compartmentation 9.8, 9.35
Definition appendix E
Fire protection of lift installations 6.41
Fire safety measures 0.16
Space separation 14.18
Auditoria
See Assembly
Automated storage systems
See Storage and other non-residential
Automatic doors
See Doors and doorways, Fire doors
Automatic fire dampers
Openings in cavity barriers 10.14
**Automatic fire detection and
alarm system** 0.14
Air extraction system 19.14
Inner rooms 4.9
Maximum dimensions of
concealed spaces 10.13
Raised storage areas 8.9
Requirement B1 page 10
Storeys divided into different
occupancies 4.14
Systems
   Dwellings 1.2-1.22
   Buildings other than dwellings
   1.23-1.32
Ventilation of common escape
routes 3.23

Automatic opening vents
Definition appendix E
Escape routes for flats or maisonettes
3.23, 3.26, diagrams 12-13
Automatic release mechanism
Definition appendix E
Holding open self-closing fire doors
appendix B(3)
Automatic self-closing devices
Definition appendix E
Fire doors appendix B(2)
Holding open fire doors appendix B(3)

B

Back gardens
Ground or basement storey exits
diagram 2
Means of escape in single family
houses 2.11
Balconies
Means of escape from flats and
maisonettes 3.9, 3.15, 3.20
Means of escape in dwelling
houses 2.6
Bars
Escape lighting table 9
Floor space factors table 1
Basement storeys
Car parks 12.4, 12.7, 19.17
Compartmentation 9.20
Construction of escape stairs 6.19
Counting the number of storeys
diagram C4
Definition appendix E
Escape routes clear of smoke
vents 19.12
Final exits clear of smoke
vents 6.34
Fire protection of lift installations
6.43, 6.44
Fire resistance of elements of
structure 8.3, table A2
Means of escape from buildings
other than dwellings 5.32, 5.5
   Added protection 5.24
   Capacity of a stair table 7
   Simultaneous evacuation 5.15
Means of escape from flats and
maisonettes 3.8, 3.40-3.42
Means of escape in single family
houses 2.10
Provision of fire fighting shafts
18.1-18.2, 18.4-18.5, diagram 51
Venting of heat and smoke
19.1-19.17
See also Underground accommodation
Bathrooms
Inner rooms 2.4
Smoke alarms 1.16
Beams
Fire resistance 8.2, table A1
Bedroom corridors
Limitations on travel distance table 3
Use of uninsulated glazed elements
on escape routes table A4
Bedrooms
Flats with alternative exits 3.12
Floor space factors table 1
Inner rooms 4.9
Limitations on travel distance table 3
Partitions and provision of cavity
barriers table 13
Positioning smoke alarms 1.11
Bed-sitting rooms
Floor space factors table 1
Betting offices
Floor space factors table 1

Billiard rooms
Floor space factors table 1
Bingo halls
Floor space factors table 1
Boiler rooms
Common stairs 3.43
Location of final exits 6.34
Places of special fire hazard
appendix E
Smoke alarms in dwellings 1.16
See also Plant rooms
Bolts
See Fastenings
Boundaries 14.4
Definition appendix E
Space separation for buildings
fitted with sprinklers 14.17
See also Notional boundaries,
Relevant boundaries
Bridges
Fire service vehicle access route
specification table 21
British Standards
476 appendix A(20)
476: Part 3 15.4, appendix A(6)
476: Part 4 B2.v, appendix A(7), table A6
476: Part 6 B2.v, appendix A(11), (14), (19)
476: Part 7 B2.v, 15.7, appendix A(10), (14), (19)
476: Part 8 appendix A(5), table A1, table B1
476: Part 11 B2.v, appendix A(7-8), tables A6-A7
476: Parts 20-24 appendix A(5), table A1
476: Part 22 appendix B(1), table B1
2782 appendix A(19), (20)
2782: Part 1 appendix A(16)
4514 table 15
5255 table 15
5266: Part 1 6.36
5306: Part 1 16.6
5306: Part 2 5.20, 8.9, 14.17, 18.7, 19.13, table 12
5395: Part 2 6.22
5438 appendix A(19)
5446: Part 1 1.4
5499: Part 1 6.37, appendix B(8)
5588: Part 0 B1.xix, 3.48
5588: Part 1 3.2, 3.6, 3.20
5588: Part 4 3.26, 4.24, 5.12
5588: Part 5 B1.xix, 6.8, 9.43, 18.11
5588: Part 6 B1.xxii, 4.18, 6.30, 6.47
5588: Part 7 0.16, 1.31, 9.8, 9.35, 14.18
5588: Part 8 B1.xvi, 6.39
5588: Part 9 6.46, 6.49, 10.13
5588: Part 10 0.16, B1.xxi, 9.20, 12.9, 18.6, table 3
5588: Part 11 B1.xxi, 4.1, 4.21, 5.5, 5.23
5720 6.49
5839: Part 1 1.3, 1.6, 1.16, 1.27, 1.29, 1.30, 5.20, 8.9, 10.13, 19.14
5839: Part 2 1.27
5839: Part 6 1.3, 1.4, 1.7, 1.16
5839: Part 8 1.28, 5.20
5867: Part 2 appendix A(19)
5906 6.50
6073 table A6
6336 appendix A(20)
6387 6.38
6661 7.8
7157 7.9
7346: Part 2 12.7, 19.14
8214 appendix B(11)
8313 9.42
CP 1007 6.36
DD 240 0.11
PD 6520 appendix A(20)
Building footprint
See Perimeter of building

Buildings of architectural or historical interest
Fire safety 0.12
Bulkheads
Smoke outlet shafts and ducts 19.15
Bungalows
See Sheltered housing, Dwelling houses
By-passing of compartmentation
See Compartments

C

Cables
Fire-stopping 11.12
Openings in cavity barriers 10.14
Power supply
smoke alarms 1.17-1.22
Protected circuits 6.38
Canopies 14.11, diagram 45
Capacity
See Occupant capacity
Car parks
Common stairs 3.43
Enclosed car parks
Fire protection of lift
installations 6.43
Escape lighting table 9
Floor space factors table 1
Internal fire spread 12.2-12.7
Maximum dimensions of building
or compartment table 12
Non-combustible materials table A6
Open sided car parks 12.4
Height of building or
compartment 14.20
Separation distances table 16
Single stair buildings 3.19
Smoke venting from basement 19.17
Cavity barriers
Cavity walls excluded diagram 32
Concealed floor or roof spaces 7.6
Construction and fixings 10.6-10.9
Definition appendix E
Double skinned insulated roof
sheeting diagram 36
Extensive ceiling voids 10.12
Fire doors table B1
Maximum dimensions 10.10-10.13
Openings 10.14, 11.5
Provision 10.2-10.5, diagram 31, table 13
Cavity walls
Excluded from provisions for
cavity barriers diagram 32
See also Insulation, Rain
screen cladding
Ceilings
Concealed spaces 10.1
Surfaces exposed 7.16
Definition appendix E
Enclosure of corridors by
partitions 4.22
Extensive concealed spaces
10.12-10.13
Fire-resisting ceilings
Concealed spaces 7.6, 10.11,
diagram 35, table 13
Fire-protecting suspended
ceilings 7.5, table A3
Lighting diffusers in suspended
ceilings 7.11, appendix A(18),
table 11
Lighting diffusers that form part
of a ceiling 7.14-7.16, diagram 23
Limited combustibility materials
table A7
Linings B2.i, 7.1
Definition 7.2-7.3
Non-combustible materials table A6
Suspended or stretched skin 7.17

Central alarm monitoring
Sheltered housing 1.10
Central cores
See Exits
Central handrails
Width of escape stairs 5.6, 5.7
Chimneys
Openings in compartment walls
or floors 9.35
Chutes
See Refuse chutes
Circulation spaces
Definition appendix E
Glazing external windows with
thermoplastic materials 7.12
Positioning smoke alarms 1.11
Separating circulation routes from
stairways 4.12
Wall and ceiling linings B2.i, table 10
Cladding
Elements of structure B3.iii
External wall construction 13.7
Overcladding 10.1
Rain screen cladding
Concealed spaces 10.1, 10.11
Provision of cavity barriers table 13
Surface of the outer cladding
facing the cavity 13.6
Class 0
Definition appendix E
Performance ratings of some generic
materials and products table A8
Product performance appendix A(12)
Use of materials of limited
combustibility table A7
Cloakrooms
Protected stairways 5.29
Clubs
Floor space factors table 1
Collapse, Resistance to B3.ii,
appendix A(5)
Columns
Fire resistance 8.2, table A1
Portal frames 13.4
**Combustible external surface
materials** 14.1
Unprotected area 14.9, diagram 43
Commercial
See Shops and commercial
purpose group
Committee rooms
Floor space factors table 1
Common balconies
Definition appendix E
Means of escape from flats and
maisonettes 3.9, 3.15
Common corridors
Escape routes for flats and
maisonettes 3.18, 3.22
Escape lighting table 9
Ventilation 3.23
Storeys divided into different
occupancies 4.14
Subdivision of common escape
routes 3.24- 3.25
Common escape routes
See Escape routes, Common
corridors
Common loadbearing elements
See Elements of structure
Common rooms
Floor space factors table 1
Common stairs
Definition appendix E
Escape lighting table 9
Flats and maisonettes 3.18,
3.29- 3.34, 3.43, diagrams 12-14
Compartment floors diagram 27
Construction 9.21-9.32
Openings 9.35
Definition appendix E
Flues diagram 39

Illustration of guidance diagram 26
Nominal internal diameter of pipes
passing through table 15
Phased evacuation 5.20
Provision 9.9-9.20
Compartment walls diagram 27
Between dwelling and corridor 3.22
Construction
Junctions 9.27-9.31, diagram 28
Openings 9.33-9.35, appendix B(5)
Definition appendix E
Elements of structure B3.iii
Flues diagram 39
Nominal internal diameter of pipes
passing through table 15
Provision 9.9-9.20
Unsuitability of cavity barriers 10.5
Compartments 9.1-9.43, diagram 27
Application of purpose groups
appendix D(1)
Basements
Direct access to venting 19.3
Calculating acceptable unprotected
area for compartments 14.20, table 16
Definition appendix E
Escape from flats and maisonettes 3.3
Fire doors table B1
Fire protection of lift installations 6.42
Flues etc 11.11
Ventilation or air conditioning
ducts 11.10
Maximum dimensions (non-residential
buildings) table 12
Measuring areas diagram C2
Progressive horizontal evacuation
4.31, 9.18
Shopping complexes 12.8
Space separation 14.2, 14.3, 14.10
Composite products
Materials of limited combustibility
appendix A(13)
Concealed spaces (cavities) 10.1
Cavity walls excluded from provisions
for cavity barriers diagram 32
Construction of cavity barriers
10.6-10.9
Definition appendix E
Fire-resisting ceilings 7.6, diagram 35
Interrupting diagram 31
Maximum dimensions 10.10-10.13
Non-residential buildings table 14
Openings in cavity barriers 10.14
Provision of cavity barriers 10.2-10.5
Roof space over protected stairway
in house with a floor over 4.5m above
ground level diagram 33
Concourses
Floor space factors table 1
Conduits for cables
Openings in cavity barriers 10.14
Openings passing through a
separating element 11.12
Conference rooms
Floor space factors table 1
Means of escape B1.xxii
Conservatories
Measuring floor area diagram C2
Plastic roofs tables 18-19
**Construction (Health, Safety and
Welfare) Regulations 1996** page 6
Corridor access
Alternative exits from flats and
maisonettes 3.15
Definition appendix E
Escape routes for flats and
maisonettes diagrams 12-13
Vision panels in doors 6.17
Corridors 4.21-4.24
Door opening 6.16
Enclosures diagram 34
Provision of cavity barriers table 13

Escape lighting table 9
Fire doors table B1
Glazed screen to protected shafts
9.39-9.40, diagram 30
Subdivision 4.22-4.24
Dead ends diagram 19
Use of uninsulated glazed elements
on escape routes table A4
Vision panels in doors 6.17
See also Bedroom corridors,
Common corridors,
Protected corridors
Courtyards
Means of escape in dwelling houses
2.11, diagram 2
Cover moulds
Definition of wall and ceiling 7.2-7.3
Crush halls
Floor space factors table 1
Cubic capacity
Measurement diagram C1
Cupboards
Fire doors appendix B(2), B(8)
Protected stairways 5.29
Curtain walls
Elements of structure B3.iii

D

Dance floors/halls
Floor space factors table 1
Dead end
Alternative means of escape B1.xiii
Definition appendix E
Fire door subdividing dead end
portions of corridor from remainder
of corridor table B1
Fire service vehicle access route
specification 17.11, diagram 50
Horizontal escape from buildings
other than dwellings 4.3
Corridors 4.21, 4.24, diagram 19
Travel distance diagram 15
Number of escape routes from flats
and maisonettes 3.18
Portion of common corridor 3.25
Use of uninsulated glazed elements
on escape routes table A4
Ventilation of escape routes from
flats and maisonettes 3.23
Deck access
Means of escape from flats and
maisonettes 3.20
Department of Health – Firecode
See Firecode
Different occupancies/uses
See Mixed use, Separate occupancies
Dining rooms
Floor space factors table 1
Direct distances
Definition appendix E
Limitations on travel distance table 3
Disabled people 6.28, 6.30
Means of escape B1.xvi
Evacuation lifts B1.xii, B1.xvi, 6.39
Phased evacuation 5.18
Travel distance limitations table 3
Vision panels in doors 6.17
Width of escape routes and exits 4.16
Discounting of stairs 5.2, 5.11-5.13
Added protection 5.24
Discounting of storey exits 4.19
Domestic garages
Construction protecting houses
9.5, 9.14, diagram 25, table A2
Fire doors between dwelling and
garage table B1
Measuring floor area diagram C2
Smoke alarms 1.16

Doors and doorways
Common escape routes 3.24- 3.26,
diagram 14
Definition of wall and ceiling
7.2-7.3, 7.12
Escape routes 6.10
Exit signs 6.37
External
Means of escape in dwelling
houses 2.11
External escape stairs 6.25
Fastening 6.11- 6.13
Fire resistance 6.6, appendix B(1)
Fire safety signs appendix B(8)
Height of escape routes 4.15, 6.26
Loft conversions 2.19
Measuring width B1.xxviii, diagram 1
Opening 6.14- 6.16
Openings in cavity barriers 10.14
Openings in compartment walls or
floors 9.33-9.35
Openings in enclosure to protected
shaft 9.43
Pressurized escape routes 4.24
Separation of circulation routes
from stairways 4.12
Subdivision of corridors 4.23
Vision panels 6.17
See also Fire doors
Dormer windows
Means of escape in
dwellinghouses 2.11
Loft conversions diagram 6
Dormitories
Floor space factors table 1
Drainage
Enclosure of pipes diagram 38
Maximum nominal internal diameter
of pipes 11.8, table 15
Dressing rooms
Inner rooms 2.4
Dry mains
See Fire mains
Ducts
Construction of smoke vent
ducts 19.15
Fire doors appendix B(2), B(8)
Openings in cavity barriers 10.14
Openings in compartment walls
or floors 9.35
Containing flues or for appliance
ventilation 11.11
Ventilation and air conditioning
11.10
Protected shafts 9.37
Ventilating ducts 9.41, 9.43
Ventilation of car parks 12.7
Dwelling houses
Calculating acceptable unprotected
area 14.15, 14.19, diagram 46
Fire doors table B1
Means of escape 2.1-2.26
Protecting from attached or integral
garages 9.5, 9.14, diagram 25, table A2
Provision of cavity barriers table 13
Provision of smoke outlets from
basements 19.4
Purpose groups table D1
Roof space over protected stairway
in house with a floor over 4.5m above
ground level diagram 33
Separation distances for roofs 15.5,
table 17
Terrace and semi-detached B3
Requirement page 60
Compartmentation 9.13
Use of uninsulated glazed elements
on escape routes table A4

Dwellings
Definition appendix E
Exit signs 6.37
Mixed use buildings 3.47- 3.48
Smoke alarms 1.2-1.22
See also Flats and maisonettes,
Residential use, Single family houses

E

Eaves
Position of emergency egress
window diagram 6
Roof measurements diagram C2
EC mark
Limitation on requirements page 5
Educational buildings
Means of escape B1.xxiii, 4.5,
5.5, table 3, table 5
Electrical circuits
See Protected power circuits
Electrical wiring
Concealed spaces 10.13-10.14
Openings passing through a
separating element 11.12
Smoke alarms 1.17-1.22
Electricity generator rooms
Escape lighting table 9
Elements of structure
Definition appendix E
Fire resistance B3.iii, 8.1-8.11,
13.1, appendix A(5)
Tests by element table A1
Emergency control rooms
Escape lighting table 9
Emergency egress windows
See Windows
Emergency lighting 6.36
Definition appendix E
Enclosed car parks
See Car parks
Enclosed space
Ground or basement storey exits
diagram 2
Enclosing rectangle
Space separation 14.15
Enclosure
Common stairs 3.33- 3.34
Corridors diagram 34
Provision of cavity barriers table 13
Drainage or water supply pipes
diagram 38
Escape stairs 5.23
Fire resistance 6.3
Enquiry office
Protected stairways 5.29
Entrance halls
See Protected entrance halls
Escalators
Means of escape B1.xii
Protected shafts 9.37
Escape lighting 6.36, table 9
Definition appendix E
Escape routes
Alternative escape and provision of
cavity barriers table 13
Clear of building 6.32
Common parts of flats and
maisonettes 3.18- 3.46
Common escape route in small
single stair building diagram 14
Limitations on distance of travel
table 2
Definition appendix E
Doors on 6.10- 6.18
Enclosure of existing stair 2.18
Fire doors table B1
Floors 6.27
Height 4.15, 6.26
Helical stairs, spiral stairs and fixed
ladders 6.22- 6.23

Horizontal escape from buildings
other than dwellings 4.1-4.32, table 3
Alternative routes diagram 16
Minimum number table 4
Width table 5
Houses with floor more than 4.5m
above ground level 2.13-2.14
Lifts prejudicing 6.40
Lighting 6.36, table 9
Measuring width B1.xxviii
Mechanical ventilation 6.46
Ramps and sloping floors 6.28- 6.30
Single steps 6.21
Siting access to refuse storage
chambers 6.53
Siting of smoke vents 19.12
Use of uninsulated glazed
elements on escape routes table A4
Width 4.16
Width relative to that of final exit 6.31
See also Protected escape routes
Escape stairs
See Stairs
Escape without external assistance
B1.i
European Technical Approval
Limitation on requirements page 5
Evacuation
Escape stairs 5.9-5.21
Added protection for phased
evacuation 5.24
Capacity of stairs for basements
and for simultaneous evacuation
table 7
Width of stairs designed for
phased evacuation table 8
Purpose of providing structure with
fire resistance 8.1
Relationships to security B1.xvii
See also Disabled people,
Progressive horizontal evacuation
Evacuation lifts
Definition appendix E
Disabled B1.xii, B1.xvi, 6.39
Exhibition halls
Floor space factors table 1
Existing buildings
Fire safety 0.12
See also Buildings of architectural
or historic interest
Exit signs 6.37
Exits
Central cores 4.10, diagram 18
Discounting of exits 4.19
Mechanical ventilation 6.46
See also Alternative exits,
Final exits, Storey exits
External assistance B1.i
External fire spread
Performance B4 guidance page 85
Requirement B4 page 84
External stair
See Stairs
External surfaces of external walls
See External walls
External walls
Concealed spaces 10.1, 10.11
Construction 13.1-13.7
Definition appendix E
Elements of structure B3.iii
External escape stairs 6.25, diagram 22
External fire spread B4.i
External surfaces 13.5, diagram 40
Facing a boundary 14.4
Fire resistance 13.3, table A1, table A2
Protected shafts 9.43
Protected stairways 3.38, 5.30, 6.24,
diagram 21
Unprotected areas 14.7-14.10,
14.13-14.16, 14.19-14.20, diagram 43,
diagram 44

F

Factory production areas
Floor space factors table 1
Travel distance table 3 note 6
Fastenings
Doors on escape routes 6.11
See also Security fastenings
Feature lifts
Fire protection of lift installations 6.41
Final exits B1.xiv, 6.31- 6.34
Definition appendix E
Enclosure of existing stair 2.18
Houses with a floor more than 4.5m
above ground level diagram 3
Protected stairways 3.35, 5.26
Siting access to refuse storage
chambers 6.53
Fire alarms 1.1-1.32
Phased evacuation 5.18
See also Automatic fire detection
and alarm system, Smoke and
fumes, Warning
Firecode (Department of Health) 0.17
Means of escape B1.xx, 4.29-4.30
Travel distances table 3
Fire doors 11.4, appendix B, table B1
Car parks 12.3
Definition appendix E
See also Doors and doorways
Firefighting lifts 18.1-18.5, 18.11
Approach 18.9, 18.12
Components of firefighting shaft
diagram 52
Definition appendix E
Flats and maisonettes 18.12
Firefighting lobbies
Approach to firefighting stair or
lift 18.9, 18.12
Components of firefighting shaft
diagram 52
Definition appendix E
Fire fighting facilities 18.1
Flats and maisonettes 18.12
Outlets from fire mains 16.5
Firefighting shafts 5.23
Components diagram 52
Definition appendix E
Design and construction 18.9-18.12
External walls
Fire resistance 9.43
Flats and maisonettes 18.1, 18.12
Glazed screens 9.39
Number and location 18.7-18.8
Building fitted with sprinklers
table 22
Provision 18.1-18.6, diagram 51
Provision of fire mains 16.2, 16.4, 18.10
Firefighting stairs 3.32, 5.3, 5.29,
6.19, 18.1
Approach through firefighting
lobby 18.9
Components of firefighting shaft
diagram 52
Definition appendix E
Fire load density
appendix A(3)
Fire mains 16.1-16.6
Fire fighting shafts 16.2, 17.6-17.7,
18.10,diagram 52
Implications for vehicle access
17.1, 17.6-17.7
Fire penetration B3.ii, appendix A(5)
Fireplace surrounds
Definition of wall 7.2
Fire Precautions Act 1971 B1.ii,
B1.iii, 1.24
Management of premises B1.vii
Powers to require staff training 0.14
Travel distances (industrial &
storage buildings) table 3

Fire Precautions (Workplace)
Regulations 1997 0.14, B1.ii, B1.iii, 1.24
Fire propagation indices
 Internal linings appendix A(11), A(12)
Fire resistance appendix A(3)-A(5)
 Definition appendix E
 Discounting radiation through
 external walls 14.2
 Elements of structure B3.iii, 8.2-8.3
 Tests by element table A1
 Minimum periods by building type
 table A2
Fire-resisting construction B1.xv,
 B3.iii, 8.1-8.3
 Alternative escape routes 4.8
 Ceilings 7.15, 7.17
 Concealed spaces 7.6, 10.11,
 table 13, diagram 35
 Compartmentation 9.1, 9.22
 Corridors 4.21
 Dead ends diagram 19
 Cupboards 5.29
 Ducts
 Openings in cavity barriers 10.14
 External escape stairs 6.25,
 diagram 22
 Fire protection of lift installations 6.42
 Flat roof forming an escape route 6.35
 Places of special fire hazard 9.12
 Protection of common stairs
 3.31- 3.39
 Refuse chutes and rooms for refuse
 storage 6.51
 Screens for subdivision of common
 escape routes 3.24- 3.25, 4.23
 Separating top storey in house from
 lower storeys 2.13, diagram 4
 Separation between garage and
 dwellinghouse diagram 25
 Shop store rooms 6.54
 Smoke outlet shafts 19.15-19.16,
 diagram 53
Fire safety engineering 0.11-0.15
Fire safety signs
 Exits 6.37
 Fire doors appendix B(8)
Fire separating elements B3.iii
 Definition Appendix E
 Fire-stopping 11.1-11.2, 11.12
Fire separation
 Adjoining protected stairways
 3.36, 5.28
 Between buildings 0.14, 9.1, 9.23
 Enclosure to protected shaft 9.43
 External walls B4.i, 14.6, table A2
 Garages and houses 9.14,
 diagram 25, table A2
 Fire resistance B3.ii-B3.iii
 Loft conversions 2.22, diagram 5
 Open sided car parks 12.4
 Openings 11.1-11.2
 Compartment walls 9.33, 9.35
 Separated parts
 Compartmentation 9.5, 9.24
 See also Cavity barriers
Fire service facilities 0.14
 Access to buildings for fire-fighting
 personnel 18.1-18.13
 Buildings not fitted with fire
 mains 17.2-17.5
 Fire mains 16.1-16.6
 Loft conversions 2.25
 Performance B5 guidance page 99
 Requirement B5 page 98
 Vehicle access 16.1,17.1
 Access routes and hardstandings
 17.8-17.11, table 21
 Blocks of flats/maisonettes 17.3
 Buildings fitted with fire mains
 17.6-17.7

 Buildings not fitted with fire mains
 17.2-17.5, table 20
 Design 17.8-17.11
 Example of building footprint and
 perimeter diagram 48
 Examples of high reach fire appliance
 access to buildings diagram 49
 Turning facilities 17.11, diagram 50
Fire spread
 Implications for means of escape B1.ix
Fire stops 11.2,11.12-11.14
 Definition appendix E
 Junction of compartment wall with
 roof 9.28, diagram 28
 Junction with slates, tiles, corrugated
 sheeting or similar 10.8
 Reinforcing materials 11.13, table A7
Fire tests
 For fire resistance by element table A1
 Methods 0.7, B2.v, appendix A(20)
 Purpose appendix A(2)
Fittings
 Fire spread B2.iv
Fixed ladders
 See Ladders
Fixed seating 6.30
 Means of escape B1.xxii
 Measuring travel distance B1.xxvii
 Spacing 4.18
Fixings
 Cavity barriers 10.9
Flame spread
 Internal linings B2.i, appendix A(9)
Flat roofs
 Means of escape from buildings
 other than dwellings 4.28, 6.35
 Means of escape from flats and
 maisonettes 3.9, 3.15, 3.28, 6.35
 Means of escape in single family
 houses 2.5-2.6
 Measuring height diagram C3
 Notional designations of coverings
 table A5
Flats and maisonettes
 Ancillary use appendix D(3)
 Calculating acceptable unprotected
 area 14.15, 14.19-14.20, diagram 46
 Compartmentation 9.15
 Conversion to 8.10
 Fire doors table B1
 Fire service facilities 18.1
 firefighting shafts 18.12
 vehicle access 17.3
 Means of escape 3.1- 3.48
 Distance of travel in common
 areas table 2
 Internal planning of flats 3.10- 3.12,
 diagram 7-9
 Internal planning of maisonettes
 3.13- 3.14, diagram 10-11
 Served by one common stair
 diagram 12
 Served by more than one common
 stair diagram 13
 Small single stair building
 diagram 14
 Provision of cavity barriers table 13
 Purpose groups table D1
 Smoke alarms 1.9
 Use of uninsulated glazed elements
 on escape routes table A4
Flexible membranes 7.9
Floor area
 Methods of measurement diagram C2
Floor level
 Change in relation to door swing 6.15
Floor space factors table 1
Floors
 Concealed spaces 10.11
 Elements of structure B3.iii, 8.4
 Escape routes 6.27

 Fire resistance 8.2, Table A1
 Fire spread by upper surfaces B2.ii
 Loft conversions 8.7
 More than 4.5m above ground level
 Means of escape from flats and
 maisonettes 3.11- 3.15
 Means of escape from houses
 2.13-2.14
 Sloping 6.29
 See also Compartment floors
Flues
 Non-combustible materials table A6
 Openings in compartment walls or
 floors 9.35, 11.11, diagram 39
Foundations
 Steel portal frames 13.4
Fuel storage space
 Common stairs 3.43
Fully supported material
 Designation of roof coverings table A5
Furnishings
 Fire spread B1.ix
Furniture
 Fire spread B2.iv
 Fitted -Definition of wall 7.2
Fusible links
 Holding open self-closing fire doors
 appendix B(3)

G

Galleries
 Counting storeys diagram C4
 Definition appendix E
 Fire resistance 8.2, table A2
 Raised storage areas 8.8
Gangways
 Between fixed storage racking table 5
 Fixed seating 6.30
 Measuring travel distance B1.xxvii
Garages
 See Domestic garages, Car parks
Gardens
 See Back gardens
Gas appliances
 Protected stairways 5.29
Gas meters
 Protected stairways 3.39, 5.31
Gas Safety (Installation & Use)
Regulations 1998 3.39, 9.41
Gas service pipes
 Protected shafts 9.41
 Ventilation 9.42
 Protected stairways 3.39, 5.31
Glazed screens
 Construction of protected shafts
 9.38-9.40, diagram 30
 Use on escape routes table A4
Glazing
 Definition of wall and ceiling 7.2-7.3
 Fire resistance of doors 2.19, 6.69
 Fire resistance of glazed
 elements 6.7- 6.9
 Glass and glazing elements on
 escape routes 6.4, table A4
 In door not part of wall 7.12
 Limitations on areas of uninsulated
 glazing table A4
 Loft conversions 2.20
 Protected shafts 9.38
 Safety 6.9
 Thermoplastic materials 7.12
 Unwired glass in rooflights 15.8
Gradient
 Hardstandings diagram 49
**Group Homes for mentally
impaired/mentally ill**
 Means of escape B1.xx
Guarding
 Flat roof forming escape route
 2.6, 6.35
 Route clear of the building 6.32

H

Habitable rooms
Definition appendix E
Escape from
flats and maisonettes 3.11- 3.14
houses 2.7, 2.8, 2.10
loft conversions 2.17
Floors in loft conversions 8.7
Hammerhead
Fire service vehicle access route
specification 17.11
Handicapped
See Disabled people
Hardstandings
Fire fighting vehicles 17.1,17.8-17.10,
diagram 49
Hardware
Fire doors appendix B(12)
See also Fastenings, Hinges
Hazard
See Places of special fire hazard, Risk
Headroom
Escape routes – general provisions
common to buildings other than
dwelling houses 6.26
Horizontal escape from buildings
other than dwellings 4.15
Health and safety
Certification of specialised
premises B1.iv
Limitation on requirements page 5
Health and Safety at Work etc Act 1974
Certification of specialised
premises B1.iv
Management of premises B1.vii
Health care
See Residential (institutional) purpose
Height
Definition appendix E
Methods of measurement
of building diagram C3
of top storey diagram C5
Helical stairs
Escape routes 6.22- 6.23
High reach appliances
Fire fighting vehicle access 17.1
Access route specification table 21
diagram 49
Buildings not fitted with fire mains
table 20
Overhead obstructions 17.10 diagram 49
High risk
See Places of special fire hazard
Hinges
Fire doors appendix B(7)
Historic buildings 0.12
Homes
See Group Homes for mentally
impaired/mentally ill, Residential
(institutional) purpose, Nursing homes
Homes for the elderly
Means of escape B1.xx
Horizontal escape
Buildings other than dwellings
4.1-4.32
Hoses
Fire mains 16.1
Hospitals
Compartmentation 4.31-4.32,
9.16-18, 9.32
HTM 81 0.17, B1.xx
Means of escape B1.xx, 4.29-4.32
Single escape routes and exits 4.5
Travel distance table 3
House conversions
See Loft conversions
Conversion to flats 8.10
Houses in multiple occupation
Means of escape B1.v, 2.3, 3.5

Houses, Single family
See Dwelling houses
Housing Act 1985
Means of escape from houses in
multiple occupation B1.v, 2.3,
Hydraulic lifts
Pipes for oil or gas in protected
shafts 9.41
Hydraulic platforms
See High Reach appliances

I

Industrial purpose group
Automatic fire detection and alarm
systems 1.31
Compartmentation 9.20
Escape lighting table 9
Escape route design
Travel distance limitations table 3
Maximum dimensions of building or
compartment table 12
Provision of cavity barriers table 13
Purpose groups table D1
Space separation
Acceptable unprotected areas
14.16
Permitted unprotected areas in
small buildings or compartments
table 16
Roofs table 17, diagram 47
See also Storage and warehousing
Inner rooms
Definition appendix E
Escape from flats and maisonettes 3.7
Escape in single family houses 2.4
Horizontal escape from buildings other
than dwellings 4.9, 4.13
In-patient care
See Hospitals, Residential
(institutional) purpose
Institutional premises
See Residential (institutional) purpose
Insulated roof sheeting
Concealed spaces 10.11
Cavity barriers diagram 36
Insulating core panels
B2.iii, B5.iii, appendix F
Insulation
Above fire protecting suspended
ceilings table A3, table A7
Concealed spaces
Flame spread rating of pipe
insulation 10.13
Space containing combustible
insulation 10.11
External wall construction 13.7
High temperatures B3.ii, appendix A(5)
Integrity
Resistance to fire penetration B3.ii,
appendix A(5)
Intercom system
Phased evacuation 5.20
Internal fire spread (linings)
Performance B2 guidance page 56
Requirement B2 page 55
Internal fire spread (structure)
Performance B3 guidance page 61
Requirement B3 page 60
Internal linings
Classification 7.1-7.17, table 10
Performance of materials
appendix A(9)-A(15)
Protection of substrate appendix A(17)
Internal speech communication system
Phased evacuation 5.20

J

Junctions
Compartmentation 9.6
Compartment wall or floor with
other walls 9.27
Compartment wall with roof
9.28-9.31, diagram 28

K

Keys
See Security fastenings
Kitchens
Fire protection of lift installations 6.43
Floor space factors table 1
Inner rooms 2.4
Means of escape from flats and
maisonettes 3.11, diagram 8
Smoke alarms 1.16
Positioning smoke alarms 1.11

L

Ladders
Means of escape B1.xii, 6.22- 6.23
Flats and maisonettes 3.3
Loft conversions 2.23, 2.25
Non-combustible materials table A6
See also High Reach appliances
Landings
Door opening and effect on escape
routes 6.15
Fire resistance of areas adjacent to
external stairs diagram 22
Floorings 6.27
See also Protected landings
Large and complex buildings
Fire safety engineering 0.11
Mixture of uses appendix D(3)
Large dwellings
Smoke alarms 1.5-1.7
Latches
See Fastenings
Laundries
See Utility rooms
Lecture halls
Means of escape B1.xxii
Libraries
Floor space factors table 1
Lift wells
Fire protection of lift installations 6.41
Use of space within protected
stairways 3.37, 5.29
Lifts
Dwellings 2.16
Exits in central core 4.10
Fire doors forming part of enclosure
of lift shaft table B1
Fire protection of installations 6.40- 6.45
Lift doors appendix B(1)
Protected shafts 9.37
Containing pipes 9.41
Openings in enclosure 9.43
See also Evacuation lifts, Fire
fighting lifts
Lighting
Diffusers forming part of a ceiling
7.14-7.16, diagram 23
Diffusers in suspended ceilings
7.11, appendix A(18), table 11
Diffusers TP(b)
Layout restrictions diagram 24
Escape routes 6.36, table 9
Lighting bridges
See Galleries
Limitations on requirements page 5
Limited combustibility
See Materials of limited combustibility
Linings
See Internal linings

Living rooms
Positioning smoke alarms 1.11
Loadbearing capacity appendix A(5)
Resistance to collapse B3.ii
Loadbearing elements of structure
See Elements of structure
Lobbies
Common escape route in small single
stair building diagram 14
Discounting of stairs 5.12, 5.13
Exits in central core 4.10
Glazed screen to protected shafts
9.39-9.40, diagram 30
Use of uninsulated glazed elements
on escape routes table A4
Ventilation of common escape
routes 3.23
See also Fire fighting lobbies,
Protected lobbies
Locks
See Fastenings
Loft conversions
Fire separation 2.22, diagram 5
Floors 8.7
Means of escape 2.17-2.25
Smoke alarms 1.8, 2.26
Lounges
Fire protection of lift installations 6.43
Floor space factors table 1
LPG see Mixed use

M

Main use
Purpose groups appendix D(1)
Maintenance
Fire safety 0.14, 0.20
Smoke alarms 1.15, 1.16
Maisonettes
See Flats and maisonettes
Malls
Application of requirements 12.8
Fire protection of lift installations 6.41
Floor space factors in shopping malls
table 1
Travel distances in shopping malls
table 3
Management
Evacuation of disabled people B1.xvi
Premises 0.14, B1.vii
See also Central monitoring
Manhole covers
Design of access routes and
hardstandings 17.8
Mantleshelves
Definition of wall 7.2
Material alteration 0.19
Materials and workmanship
Independent certification schemes
page 5
Regulation 7 page 5
Materials of limited combustibility
Class 0 appendix A(12)
Composite products appendix A(13)
Concealed spaces 10.13
Construction of escape stairs 6.19
Definition appendix E
Insulation above fire protecting
suspended ceilings table A3
Insulation materials used in
external wall construction 13.7
Junction of compartment wall
with roof 9.29, diagram 28
Reinforcing materials used for
fire-stopping 11.13
Roofs table 17
Use appendix A(8), table A7
Means of escape
Compartment walls 4.31, 9.18
Openings 9.33
Criteria B1.xi-B1.xii
Definition appendix E

Performance B1 guidance page 11
Protected shafts
Openings in enclosure that is
wall common to two buildings 9.43
Requirement B1 page 10
Measurement
See Methods of measurement
Mechanical ventilation
See Ventilation
Mechanised walkways
Means of escape B1.xii
Meeting rooms
Floor space factors table 1
Methods of measurement appendix C
Means of escape
Occupant capacity B1.xxvi
Travel distance B1.xxvii
Width of escape routes, doors and
stairs B1.xxviii
Mixed use
Car park and other use 12.4
Compartment walls and
compartment floors 9.11
Division of storeys 9.13
Dwellings in mixed use buildings
3.47, 3.48
Escape stairs 5.2, 5.4, 5.5
Horizontal escape 4.4
Storage of petrol & LPG 3.48
Multiple occupation
See Houses in multiple occupation
Multi-storey buildings
Horizontal escape from buildings
other than dwellings 4.3, 4.10
Vertical escape from buildings other
than dwellings 5.1

N

Natural ventilation
See Ventilation
Non-combustible materials
Definition appendix E
Sleeving non-combustible pipes 11.9
Use appendix A(7), table A6
Non-residential buildings
Calculating acceptable unprotected
area 14.20, table 16
Compartmentation 9.20
Maximum dimensions of concealed
spaces table 14
Non-residential buildings
Maximum dimensions of building
or compartment table 12
Purpose groups table D1
Wall and ceiling linings 7.4, table 10
Notional boundaries 14.6, diagram 42
Definition appendix E
Separation distances for roofs 15.5
Noxious gases
Fire spread B1.x
Nursing homes
Means of escape B1.xx

O

Occupancies
See Mixed use, Separate occupancies
Occupant capacity
Horizontal escape from buildings other
than dwellings 4.5, 4.7, 4.9, 4.17
Methods of measurement B1.xxvi
Vertical escape from buildings other
than dwellings
Minimum width of escape stairs
5.10, table 6
Offices
Automatic fire detection and alarm
systems 1.31
Compartmentation 9.20
Escape lighting table 9
Fire doors table B1

Floor space factors table 1
Junction of compartment wall with
roof 9.30
Maximum dimensions of building
or compartment table 12
Means of escape
Alternative approach B1.xix, 4.1
Limitation of travel distance table 3
Single escape stairs 5.5
Small premises 4.1
Provision of cavity barriers table 13
Purpose groups table D1
Space separation
Acceptable unprotected areas 14.16
Permitted unprotected areas in
small buildings or compartments
table 16
Open planning
Awareness of fires 4.22
Escape lighting table 9
Simultaneous evacuation 5.15
Open sided car parks
See Car parks
Open spatial planning
Definition appendix E
Openable vents
Escape routes for flats or
maisonettes 3.23, diagram 12,
diagram 13, diagram 14
Openable windows
See Windows
Opening characteristics
See Doors and doorways
Openings
Cavity barriers 10.14
Compartment walls or floors 9.35
Compartment walls separating
buildings or occupancies 9.33
Emergency egress in dwelling
houses 2.11
Protected shafts 9.43
Protection 11.1-11.14
See also Doors and doorways,
Windows
Other non-residential
See Storage and other non-residential
Other residential use
Calculating acceptable unprotected
area 14.15, 14.16, 14.19, 14.20,
diagram 46, table 16
Compartmentation 9.19
Fire doors table B1
Junction of compartment wall with
roof 9.30
Limitations on travel distance table 3
Progressive horizontal evacuation 4.30
Provision of cavity barriers table 13
Purpose groups table D1
Simultaneous evacuation 5.15
Outbuildings
Measuring floor area diagram C2
Overcladding 10.1
External wall construction 13.7
Overhanging storey
See Perimeter of building
Overhead obstructions
Access of High reach appliances
17.10, diagram 49

P

Partitions
Cavity barriers 10.6
Enclosing corridors 4.22
Fire resistance 6.3
Provision of cavity barriers in
bedrooms table 13
See also Walls
Pavement lights
Smoke vent outlet terminal 19.11

Performance
Access and facilities for fire service B5 page 99
External fire spread B4 page 85
Internal fire spread
Linings B2 page 56
Structure B3 page 61
Means of escape B1 page 11
Protection systems, materials and structures 0.20
Ratings of some generic materials and products table A8

Perimeter of building
Definition appendix E
Example of footprint and perimeter diagram 48
Fire service vehicle access table 20

Permanent ventilation
See Ventilation

Phased evacuation
See Evacuation

Picture rails
Definition of wall and ceiling 7.2-7.3

Pipes
Cavity barriers 10.14
Compartment walls or floors 9.33, 9.35
Maximum nominal internal diameter 11.7, table 15
Definition appendix E
Flame spread rating of insulation
Maximum dimensions of concealed spaces 10.13
Non-combustible materials table A6
Openings 11.5-11.9,11.12
Penetrating structure diagram 37
Protected shafts 9.37
Oil or gas 9.41
Openings 9.43
See also Gas service pipes

Pitched roofs
Measuring height diagram C3
Notional designations of coverings table A5

Place of safety
Enclosed space diagram 2
Inside building B1.xi
Progressive horizontal evacuation 4.31

Places of special fire hazard
Compartmentation 9.5, 9.12
Definition appendix E
Escape stairs 5.25
High risk and lift installations 6.43
Inner rooms 4.9
Limitation on travel distance table 3
Smoke vents 19.9

Plant rooms
Fixed ladders 6.22
Limitations on travel distance table 3

Plastics
Rooflights 7.7, 15.4, 15.6-15.7, table 11, table 18, table 19
Layout restrictions diagram 24
Spacing and size diagram 47

Platform floors
Definition appendix E
Exclusions from elements of structure 8.4

Plenum
Maximum dimensions of concealed spaces 10.13

Podium
External escape stairs 3.46
Fire resistance of adjacent areas diagram 22

Polycarbonate sheet
TP(a) rigid and TP(b) appendix A(19)

Porches
Smoke alarms 1.16

Portal frames 13.4

Power supply
Cables
Smoke alarms 1.17
Mechanical ventilation for car parks 12.7
Protected power circuits 6.38

Pressurization 3.26
Compatibility with ventilation and air conditioning 6.48
Corridors 4.24
Ducts in protected shafts 9.41

Private stair
See Stairs

Progressive horizontal evacuation
4.30-4.32, diagram 20

Projecting upper storey
See Overhead obstructions

Property protection 0.18

Protected corridors 4.21
Added protection for escape stairs 5.24
Dead end corridors diagram 19
Definition appendix E
Escape routes for flats and maisonettes
Ancillary accommodation 3.27
Basements 3.42
Fire doors table B1
Phased evacuation 5.20
Protection of common escape routes 3.22
Storeys divided into different occupancies 4.14
Use of uninsulated glazed elements on escape routes table A4

Protected entrance halls
Definition appendix E
Escape from flats and maisonettes 3.11, 3.14, diagram 7, diagram 11
Fire doors forming part of enclosure in flat or maisonette table B1

Protected escape routes B1.xiv
Corridors 4.21
Dead ends diagram 19
Discounting of stairs 5.11
Escape stair needing added protection 5.24
Phased evacuation 5.20
Protection and compartmentation 9.1
Protection for buildings other than dwelling houses 6.2- 6.4
Storeys divided into different occupancies 4.14

Protected landings
Alternative exits from maisonettes 3.14
Definition appendix E
Fire doors in flats or maisonettes table B1

Protected lobbies
Ancillary accommodation 3.27, 3.44
Definition appendix E
Discounting of stairs 5.12
Escape stairs 5.24-5.25
Fire doors table B1
Fire protection of lift installations 6.42- 6.43
Number of escape routes 3.18
Phased evacuation 5.20
Planning of common escape routes 3.21
Refuse chutes and rooms for refuse storage 6.51- 6.52
Storey exits 4.11
Use of uninsulated glazed elements on escape routes table A4
Ventilation 6.52
See also Lobbies

Protected power circuits 6.38
Definition appendix E

Protected shafts 5.23, 9.7, 9.36-9.43, 18.1, diagram 29
Definition appendix E
External wall of stairs in a protected shaft 14.8, diagram 44

Fire doors table B1
Fire protection of lift installations 6.42
Glazed screens 9.39-9.40, diagram 30
Openings 9.43
Openings for pipes 11.5
Pipes for oil or gas 9.41
Pipes passing through structure enclosing protected shaft table 15
Ventilation of shafts conveying gas 9.42
Ventilating ducts 9.41
See also Firefighting shafts

Protected stairways B1.xv
Air circulation systems for heating, energy conservation or condensation control in buildings with a floor more than 4.5m above ground level
Flats and maisonettes 3.16
Houses 2.15
Common stairs 3.31- 3.39
Definition appendix E
Escape routes from common parts of flats and maisonettes 3.29- 3.39
Escape stairs 5.1-5.31
Exits in central core diagram 18
External protection diagram 21
Fire doors table B1
Fire protection of lift installations 6.40- 6.41, 6.45
Houses with floor more than 4.5m above ground level 2.13
Protected shafts 9.36-9.37
Containing pipes 9.41
Unprotected areas 14.8
Refuse chutes and rooms for refuse storage 6.51
Roof space above in house with a floor over 4.5m above ground level diagram 33
Rooflights of thermoplastic materials 7.13
Separation of circulation routes from stairways 4.12
Thermoplastic lighting diffusers in ceilings 7.16
Use of space below diffusers or rooflight table 11
Use of uninsulated glazed elements on escape routes table A4
Vertical escape from buildings other than dwellings 5.26-5.30

Protective barriers
Flat roof forming escape route 2.6, 6.35

PTFE-based materials 7.10

Public address system 1.28

Publications (excluding BSI)
Approved Document:
F Ventilation. 12.6, 12.7
J Combustion appliances & fuel storage. diagram 39, 15.9
K Protecton from falling, collision & impact. 2.6, 2.11, 2.21, 5.7, 6.20, 6.23, 6.30, 6.35
M Acces & facilities for disabled people. 4.16, table 5, table 6, 6.17, 6.28, 6.30
N Glazing – safety in relation to impact, opening & cleaning. 6.9, 6.17, 7.12
Assessing the fire performance of external cladding systems: a test method, BRE. 13.5
Building Regulation and Fire Safety – Procedural Guidance. B1.iii
CE Marking Directive. page 5
Construction fire safety. CIS 51. page 6
Construction Products Directive. page 5
Crown Fire Standards. 0.18

Design, construction, specification and fire management of insulated envelopes for temperature controlled environments. IACSC. F(8)

Design methodologies for smoke and heat exhaust ventilation. BRE. table 3, 12.7

Draft guide to fire precautions in existing residential care premises. B1.xx

External fire spread: Building separation and boundary distances. BRE. 14.15, 14.16, 14.20

Fire and steel construction: The behaviour of steel portal frames in boundary conditions. SCI. 13.4

Firecode. B1.xx, 4.29, 4.30, table 3, 9.32, table A2
    HTM 81. 0.17, B1.xx
    HTM 88. B1.xx

Fire performance of external thermal insulation for walls of multi-storey buildings. BRE. 13.7

Fire Precautions (Workplace) Regulations 1997. 0.14, B1.ii, B1.iii, 1.24

Fire protection for structural steel in buildings. ASFP. A(5)

Fire safety in construction work, HSG 168. page 6

Fire safety of PTFE-based materials used in buildings. BRE. 7.10

Gas Safety (Installation and Use) Regulations 1998. 3.39, 9.41

Guidelines for the construction of fire-resisting structural elements. BRE. A(1)

Guide to fire precautions in existing places of entertainment and like premises. B1.xxii

Guide to fire precautions in existing places of work that require a fire certificate. Factories, offices, shops & railway premises. table 3

Guide to Safety at Sports Grounds. B1.xxii

Hardware essential to the optimum performance of fire-resisting timber doorsets. ABHM. B(12)

Heat radiation from fires and building separation. Fire Research Technical Paper 5. 14.16

Houses in multiple occupation
    DOE Circular 12/92. 2.3
    DOE Circular 12/93. 2.3
    WO Circular 25/92. 2.3
    WO Circular 55/93. 2.3

Increasing the fire resistance of existing timber floors. BRE Digest 208. table A1

LPC Design guide for the fire protection of buildings. 0.18

Pipelines Safety Regulations 1996. 3.39, 9.41

Safety signs & signals.
    Health & Safety (Safety signs & signals) Regulations 1996.
    Guidance on Regulations. 6.37

Thatched buildings. New properties and extensions. 15.9

Workplace (Health, Safety and Welfare) Regulations 1992, Approved Code of Practice and Guidance. page 6

Pumping appliances
    Dry fire mains 17.6
    Fire fighting vehicle access 17.1
        Access route specification table 21
        Blocks of flats/maisonettes 17.3
        Buildings not fitted with fire mains table 20
    Replenishing wet fire mains 16.1, 17.7

Purpose groups appendix D
    Classification table D1
    Definition appendix E
    Fire safety 0.6

PVC sheet
    TP(a) rigid appendix A(19)

Q

Quantitative techniques to evaluate risk and hazard 0.15

Queueing areas
    Floor space factors table 1

R

Racking
    See Storage and Other nonresidential

Radiation
    Basis for calculating acceptable unprotected area 14.16
    Discounting 14.2

Rafters
    Portal frames 13.4

Rain screen cladding
    Concealed spaces 10.1, 10.11
    Provision of cavity barriers table 13
    Surface of the outer cladding facing the cavity 13.6

Raised storage
    Fire resistance 8.8-8.9
    See Storage and other nonresidential

Ramps
    Escape routes 6.27-6.30

Reading rooms
    Floor space factors table 1

Reception desk
    Protected stairways 5.29

Recirculating distribution systems
    Mechanical ventilation 6.46
    Maximum dimensions of concealed spaces 10.13

Recreation
    See Assembly and recreation purpose group

Refuges
    Means of escape for disabled B1.xvi, table A4

Refuse chutes 6.50-6.52
    Non-combustible materials table A6
    Openings in compartment walls or floors 9.35
    Protected shafts 9.37

Refuse hoppers 6.50

Refuse storage chambers
    Access 6.53
    Compartment walls 9.15
    Location of final exits 6.34

Relevant boundaries 13.1,14.4-14.5, diagram 41
    Definition appendix E
    External walls 1000mm or more from relevant boundary 14.14
    External walls within 1000mm from relevant boundary 14.13
    Other method for calculating acceptable unprotected area 14.20
    Portal frame building near boundary 13.4
    Separation distances for roof 15.5, table 17
    Small residential method 14.19

Residential (institutional) purpose
    Automatic fire detection and alarm systems 1.30
    Compartmentation 9.16-9.18
    External escape stairs 5.33
    Hospitals 0.17

Means of escape B1.xx, 4.29-4.32
    Inner rooms 4.9
    Limitations on travel distance table 3
    Single escape routes and exits 4.5
Minimum width of escape stairs table 6
Provision of cavity barriers table 13
Purpose groups table D1
Use of uninsulated glazed elements on escape routes table A4

Residential use
    Automatic fire detection and alarm systems 1.30
    Escape lighting table 9
    Escape route design 4.4
        Protected corridors 4.21
    Mixed use 3.47- 3.48, 5.4
    Purpose groups table D1
    Space separation 14.2
        Acceptable unprotected areas 14.16
        Notional boundaries 14.6
        Permitted unprotected areas in small building or compartments 14.19, 14.20, table 16, diagram 46
    Wall and ceiling linings 7.4, table 10
    See also Flats and maisonettes, Other residential use, Residential (institutional) purpose, Dwelling houses

Restaurants
    Escape lighting table 9
    Floor space factors table 1

Revolving doors
    See Doors and doorways

Rising fire mains
    See Fire mains

Risk
    Insurance 0.18
    Quantitative techniques of evaluation 0.15
    See also Places of special fire hazard

Rolling shutters 18.13, appendix B(2), B(5), B(6)

Roof coverings
    Concealed spaces between insulated roof sheeting 10.11
        Cavity barriers diagram 36
    External fire spread 15.1-15.9
        Separation distances for roofs table 17
    Junction of compartment wall with roof 9.28-9.31, diagram 28
    Materials of limited combustibility table A7
    Notional designations table A5

Roof space
    Concealed space 10.1
    Over protected stairway in house with a floor over 4.5m above ground level diagram 33
    See also Loft conversions

Roof terraces
    Means of escape 2.24

Roof windows
    Emergency egress from single family houses 2.11, diagram 6

Rooflights
    Definition appendix E
    Definition of ceiling 7.3
    Loft conversions 2.24, diagram 6
    Methods of measurement diagram C2
    Plastic 7.7, 15.4, 15.6-15.7, table 11, table 18, table 19
        Layout restrictions diagram 24
        Limitations on spacing and size diagram 47
    Thermoplastic 7.11, 7.13, appendix A(18), table 11
        Layout restrictions diagram 24
    Unwired glass 15.8

Roofs 15.2
  Concealed spaces between
  insulated roof sheeting 10.11
    Cavity barriers diagram 36
  Elements of structure B3.iii, 8.4
  Escape over 2.5, 3.28, 4.28, 6.35
  External escape stairs 3.46
    Fire resistance of adjacent areas
    diagram 22
  Fire resistance 6.3, appendix A(6)
  Junction of compartment wall with
  roof 9.28-9.31, diagram 28
  Materials of limited combustibility
  table A7
  Measuring
    area diagram C2
    height diagram C3
  Rooftop plant
    Height of top storey in building C5
    Limitations on travel distance
    table 3
  See also Flat roofs, Pitched roofs
Rooms
  Definition appendix E
  Measuring floor area diagram C2

S

Safety of Sports Grounds Act 1975
B1.xxii
Sanitary accommodation
  Protected shafts 9.37
  Protected stairways 5.29
Sanitary towel incinerators
  Protected stairways 5.29
Schools
  See Educational buildings
Seals
  Fire resistance 11.2
  Proprietary seals 11.14
    Pipes 11.6
Seating
  See Fixed seating
Seatways
  Measuring travel distance B1.xxvii
Security
  Conflict with requirements for means
  of escape B1.xvii
Security fastenings
  Common escape routes diagram 14
  Doors on escape routes 6.11- 6.12
Self-closing devices
  See Automatic self-closing devices
Self-contained smoke alarms
  See Smoke
Self-supporting sheets
  Designation of pitched roof coverings
  table A5
Semi-detached houses B3 requirement
page 60
  Compartment walls 9.13
  Separation distances 15.5
Separate occupancies/uses
  Compartment walls and
  compartment floors 9.20
  Corridors 4.21
  Division of storeys 4.13, 4.14
  Openings in compartment walls 9.33
Separated parts of buildings 9.5, 9.24
  Definition appendix E
Separation
  See Fire separation, Space separation
Separation distances
  Buildings with sprinkler systems 14.17
  Canopies 14.11, diagram 45
  Compartment size 14.3
  Roofs 15.5, table 17
  Unprotected areas which may be
  disregarded diagram 44

Services
  Penetrating cavity barriers 10.9
  See also Air conditioning, Ducts,
  Pipes, Ventilation
Shafts
  Fire-resisting construction for smoke
  outlet shafts 19.15-19.16, diagram 53
  See also Fire fighting shafts,
  Protected shafts
Sheltered housing
  Central alarm monitoring 1.10
  Means of escape 3.6
Shopping complexes
  Compartmentation 9.20, 12.8
  Firefighting shafts 18.6
  Floor space factors table 1
  Internal fire spread 12.8-12.9
  Limitations on travel distance table 3
  Means of escape B1.xxi
Shop and commercial purpose group
  Automatic fire detection and alarm
  systems 1.31
  Compartmentation 9.20
  Escape lighting table 9
  Floor space factors table 1
  Limitations on travel distance table 3
  Maximum dimensions of building or
  compartment table 12
  Panic bolts 6.12
  Provision of cavity barriers table 13
  Purpose groups table D1
  Small shops see Small premises
  Space separation
    Acceptable unprotected areas 14.16
  Permitted unprotected areas
    in small buildings or
    compartments table 16
  Store rooms 6.54
Showers
  Inner rooms 2.4
  Smoke alarms 1.16
Signs
  See Exit signs, Fire safety signs
Simultaneous evacuation
  See Evacuation
Single stair buildings
  Added protection for escape stair 5.24
  Common escape route diagram 12,
  diagram 14
  Construction of escape stairs 6.19
  Continuation of lift down to
  basement storey 6.44
  Location of lift machine rooms 6.45
  Small buildings 3.19
  Ventilation of common escape
  routes 3.23
Single steps
  Escape routes 6.21
Single storey buildings
  Definition appendix E
  Fire resistance of elements of
  structure 8.3, table A2
  See also Single family houses,
  Sheltered housing
Skating rinks
  Floor space factors table 1
Skirtings
  Definition of wall 7.2
Slates or tiles
  Designation of pitched roof
  coverings table A5
  Fire stopping junctions 10.8
Sleeping accommodation
  Fire protection of lift installations 6.43
  See also Bedrooms, Dormitories
Sleeping galleries 2.9, 3.2
Sleeving non-combustible pipes 11.9
Slipperiness
  Floors of escape routes 6.25

Small premises
  Enclosure of escape stairs 5.23
  Means of escape 4.1
Smoke and fumes
  Alarms in dwellings 1.2-1.22
    Installation 1.4-1.22
  Basement smoke vents 19.1-19.17
    Fire-resisting construction for
    smoke outlet shafts diagram 53
  Ceiling level smoke vents in car
  parks 12.6
  Control systems 0.14, B1.xv, 3.26
  Discounting of stairs 5.12
    Protected lobbies 5.25
    Shopping malls table 3
  Final exits clear of basement
  smoke vents 6.34
  Fire spread B1.x
    Retarding by sealing or
    firestopping 11.3
  Fire spread and lining materials B2.i
  Lift machine room fires 6.45
  Smoke leakage of fire doors at
  ambient temperatures appendix B(1),
  table B1
  Smoke reservoirs 6.41
  Ventilation of common escape
  routes 3.23
Space separation 14.1-14.20
  Roofs 15.5, table 17
  Unprotected areas disregarded
  in assessing separation distance
  diagram 44
Special applications 7.8-7.10
Spiral stairs
  Escape routes 6.22- 6.23
Sports grounds
  Means of escape B1.xxii
Sprinkler systems 5.20
  Discounting of stairs 5.13
  Maximum dimensions of building
  or compartment
    Non-residential buildings table 12
  Mechanical smoke extract 19.13
  Number and location of firefighting
  shafts 18.7, table 22
  Portal frames 13.4
  Space separation 14.2
    Boundary distance 14.17
  Permitted unprotected areas in small
  building or compartments table 16
Stability B3 requirement page 60
Staff rooms
  Floor space factors table 1
Staff training
  Powers to require 0.14
Stage grids
  See Galleries
Stairs
  Accommodation stairs 5.23
  Alternative exits from flats and
  maisonettes 3.15
  Basement
    Means of escape 3.40- 3.42, 5.32
  Continuing beyond the level of
  final exit 6.33
  Design 6.20
  Escape stairs 5.1-5.34
    Capacity for basements and
    simultaneous evacuation table 7
    Construction 6.19
    Continuing lift down to
    basement 6.44
    Lighting 6.36
    Width table 6
    Width where designed for
    phased evacuation table 8

External escape 3.15, 3.45- 3.46, 4.26, 5.33-5.34, 6.25
    Fire resistance of adjacent areas diagram 22
    Materials of limited combustibility 6.19
Firefighting 3.32, 5.3, 5.29, 6.19, 18.1
    Approach through firefighting lobby 18.9
    Components of firefighting shaft diagram 52
Fire spread by upper surfaces B2.ii
Helical and spiral 6.22- 6.23
Horizontal escape from buildings other than dwellings 4.3
Limited combustibility materials table A7
Loft conversions 2.18, 2.21
Public stair 6.22
Serving accommodation ancillary to flats and maisonettes
    Means of escape 3.43- 3.44
Treads of steps 6.27
See also Common stairs

Stairways
Door opening 6.16
Escape lighting table 9
Measuring travel distance B1.xxvii
Measuring width B1.xxviii
Primary circulation routes 4.12
See also Protected stairways

Stallboard
Fire-resisting construction for smoke vent shafts diagram 53
Smoke vent outlet terminal 19.11

Standard fire tests
See Fire tests

Standing spectator areas
Floor space factors table 1

Steps (single)
Escape routes 6.21

Storage and other non-residential
Automatic fire detection and alarm systems 1.31
Compartmentation 9.20
Escape lighting table 9
Fire service vehicle access to buildings without fire mains table 20
Floor space factors table 1
Interpretation of purpose groups appendix D(1)
Limitations on travel distance table 3
Maximum dimensions of building or compartment table 12
Provision of cavity barriers table 13
Purpose groups table D1
Space separation
    Acceptable unprotected areas 14.16
    Permitted unprotected areas in small building or compartments table 16
    Roofs table 17
Widths of escape routes 4.16, table 5

Stores
Fire protection of lift installations 6.43
In shops 6.54
Raised storage areas 8.8-8.9
Refuse storage 6.50- 6.53
    Compartment walls 9.15

Storey exits 4.1
Access in buildings other than dwellings 4.11, 4.24
Definition appendix E
Discounting 4.19
Escape over flat roofs 6.35
Leading into an enclosed space diagram 2

Storeys
Definition appendix E
Divided into different
    occupancies 4.14
    uses 4.13
Height of top storey diagram C5
Number diagram C4

Stretched skin ceilings
Thermoplastic material 7.17

Strong rooms
Smoke vents 19.5

Structural frames
Elements of structure B3.iii
Fire resistance 8.2

Student residential accomodation
Automatic fire detection and alarm systems 1.9

Structural loadbearing elements
See Elements of structure

Stud walls
Cavity barriers 10.6

Study bedrooms
Floor space factors table 1

Substrates
Lining to wall or ceiling appendix A(17)

Surface spread of flame
Internal linings appendix A(10)-A(12), A(15)

Susceptibility to ignition
External walls 13.2

Suspended ceilings
Concealed spaces 10.1
    Wall and ceiling surfaces exposed 7.16
Definition appendix E
Enclosure of corridors by partitions 4.22
Lighting diffusers 7.14, table 11
Limited combustibility materials table A7
Non-combustible materials table A6
Thermoplastic material 7.17, appendix A(18)
Use of fire protecting suspended ceilings table A3

Switch room/battery room
Escape lighting table 9

T

Tall buildings
Added protection for escape stairs 5.24
External walls 13.2
Width of escape stairs 5.6
    Phased evacuation 5.18

Telephone system
Phased evacuation 5.20

Tenancies
See Separate occupancies

Terrace house B3 requirement page 60
Compartment walls 9.13

Theatres
Means of escape B1.xxii

Thermoplastic materials
appendix A(16)-A(19)
Classification B2.v
Definition appendix E
Glazing 7.12
Lighting diffusers forming part of a ceiling 7.14-7.16
Lighting diffusers in suspended ceilings 7.11, table 11
Rooflights 7.11, 7.13, 15.6-15.7, table 11, table 18, table 19
Suspended or stretched skin ceiling 7.17
TP(b) rooflights and lighting diffusers
    Layout restrictions diagram 24

Tiers
Pitch 6.29

Tiles
See Slates or tiles

Toilets
Escape lighting table 9
Exits in central core diagram 18

TP(a) flexible
Classification of performance B2.v, appendix A(19)

TP(a) rigid
Classification of performance B2.v, appendix A(19)
Glazing external windows 7.12

TP(b)
Classification of performance B2.v, appendix A(19)
Rooflights and lighting diffusers
    Layout restrictions diagram 24

Transfer of excessive heat
Insulation from high temperatures B3.ii, appendix A(5)

Transformer chambers
Final exits clear of openings 6.34
See also Place of special fire hazard

Trap doors
See Doors and doorways

Travel distance B1.xi
Definition appendix E
Horizontal escape from buildings other than dwellings 4.2, 4.5, 4.6, 4.9, table 3
    Dead end diagram 15
Means of escape from flats and maisonettes 3.11, 3.18, diagram 8, diagrams 12-14, table 2
Methods of measurement B1.xxvii
Vertical escape from buildings other than dwellings 5.23

Travelators
See Mechanised walkways

Treads
See Stairs

Turning circles
Fire service vehicle access route specification 17.11, table 21

Turnstiles
Escape routes 6.18

Turntable ladders
See High Reach appliances

U

UKAS
Accredited laboratories appendix A(1)
Independent certification schemes page 5

Underground accommodation
Escape lighting table 9

Uninsulated glazing
See Glazing

Unobstructed openings
Means of escape in single family houses 2.11

Unprotected areas 14.1
Areas disregarded in assessing separation distance diagram 44
Boundaries 14.4
Combustible material as external surface 14.9, diagram 43
Definition appendix E
External wall of protected shaft forming stairway 14.8
External wall within 1000mm of relevant boundary 14.13
External wall 1000mm or more from relevant boundary 14.14-14.20
    Permitted unprotected areas in small buildings or compartments 14.20, table 16
    Permitted unprotected areas in small residential buildings 14.19, diagram 46
Fire resistance 14.7

Large uncompartmented
buildings 14.12
Small unprotected areas 14.10,
diagram 44
Unwired glass
Rooflights 15.8
Utility rooms
Inner rooms 2.4

V

Vehicle access
See Fire service facilities
Ventilation
Basement smoke vents 19.1-19.17
Escape routes 19.12
Final exits 6.34
Fire-resisting construction for
smoke outlet shafts diagram 53
Car parks 12.2, 12.4-12.7, 19.17
Ducts and integrity of compartments
11.10-11.11
Ducts in protected shafts 9.41
Openings in enclosure 9.43
Escape routes for flats or maisonettes
3.23, 3.26, diagram 12, diagram 13,
diagram 14
Lobbies 5.25
Mechanical 6.46- 6.49
Concealed space used as a
plenum 10.13
Smoke extract 19.13-19.14
Openings in compartment walls
or floors 9.35
Protected shafts 9.38
Conveying gas 9.42
Ventilating ducts 9.41
Rooms containing refuse chutes
or for the storage of refuse 6.52
Venues for pop concerts
Floor space factors table 1
Verges
Roof measurements diagram C2
Vertical escape
See Stairs, Stairways
Vision panels
Doors on escape routes 6.17
Inner rooms
Buildings other than dwellings 4.9
Voice alarm system 1.28, 5.20
Volume of building or part
Measurement diagram C1

W

Waiting rooms
Floor space factors table 1
Walkways
See Mechanised walkways
Wall climber
Fire protection of lift installations 6.41
Walls
Cavity barriers in stud walls or
partitions 10.6
Cavity walls excluded diagram 32
Concealed spaces 10.11
Elements of structure B3.iii
Fire resistance 6.3, 8.2, 13.3, table A1
In common to two buildings
Calculating perimeter diagram 48
Compartmentation 9.10, 9.23
Inner rooms
Buildings other than dwellings 4.9
Linings B2.i, 7.1, 7.4
Definition 7.2, 7.12
Surfaces within the space above
suspended ceilings 7.16
See also Cavity walls, Compartment
walls, External walls

Warehouses
See Storage and other non-residential
Warning 0.14
Escape from
flats and maisonettes 3.1
houses 2.1
Phased evacuation 5.20
Smoke alarms 1.1
Washrooms
Protected shafts 9.37
Protected stairways 5.29
Water supply pipes
Enclosure diagram 38
WCs
Inner rooms 2.4
Weather protection
External escape stairs 6.25
Wet mains
See Fire mains
Wind loads
Elements of structure B3.iii
Windowless accommodation
Escape lighting table 9
Windows
Definition of wall and ceiling 7.2-7.3
Emergency egress in dwelling
houses 2.11
Loft conversions 2.23
Thermoplastic materials A(18)
Glazing 7.12
Wiring
See Electrical wiring
Workmanship
See Materials and workmanship
Workplace (Health, Safety and Welfare)
Regulations 1992 page 6
Workshops
Floor space factors table 1

692 3